Molecular Biology
Biochemistry and Biophysics

Molekularbiologie
Biochemie und Biophysik

4

Springer-Verlag New York Inc. 1969

A. S. Spirin · L. P. Gavrilova

The Ribosome

With 26 Figures

Springer-Verlag New York Inc. 1969

Professor Dr. A. S. Spirin
Dr. L. P. Gavrilova

A. N. Bakh Institute of Biochemistry and Institute of Protein Research
Academy of Sciences of the USSR, Moscow, USSR

 Library of Congress Catalog Card Number 69-20346. Printed in Germany.

Title No. 3804

Preface

This monograph is neither a historical outline of the development of the concepts of protein biosynthesis and the structure and functions of the ribosomes, nor an exhaustive survey of the literature on these questions. The monograph is based upon an analysis of the modern trends in this field. The purpose of the monograph was to formulate more or less generalized representations of the structure and function of the ribosome, as we envision it at the present day. It may be that this attempt is premature for a number of reasons, and the concepts outlined here will very soon be revised. Nonetheless, despite this risk, we believe it to be advisable to undertake this attempt for the following reasons: firstly, the undertaken analysis could aid in the comprehension of the substantial mass of extremely scattered experimental data on the ribosomes presently available; secondly, in any event, even if most of the concepts outlined rapidly become obselete, they can still serve as a stimulus for a whole series of experiments; and thirdly, we hope that some of the concepts outlined will still remain essentially correct and relatively stable.

In view of the aforementioned, we should make the following reservations. First of all, we made no attempt to cite all the literature on the problems discussed, but considered it sufficient to illustrate the various premises with one or several sample references. Of course, here we endeavored to select the clearest and most indicative works. Nonetheless, many even very lucid studies, which have played a great role in the development of modern concepts of protein biosynthesis, were not included in the list of literature cited — either because they are now to a considerable degree only of historical interest, or because they are so very well known to the scientific reader. Finally, not all the questions associated with the ribosomes are discussed in the monograph. For example, we did not at all touch upon the question of the polysomes — their distribution in the cells, structural organization, and biological significance. Nor did we take up among others the question of the attachment of the ribosomes to membranes. It is no accident that the monograph is entitled "The Ribosome", and not "Ribosomes": we made an effort to concentrate precisely upon the structural organization and mode of functioning of the *individual* ribosomal particle as such, as a complete molecular machine.

Many of the "gaps" in the monograph, enumerated and left purposely, can be filled by the available literature. A complete summary of the studies on the physical and chemical properties of the ribosomes, published up to 1964, inclusively, is given in M. Petermann's monograph (PETERMANN, 1964). The results of original investigations of the properties, structure, and especially biogenesis of ribosomes are assembled in the collection edited by R. ROBERTS (ROBERTS, 1964). A brilliant exposition of the problems of the structure and especially the function of the ribosomes can be found in J. Watson's articles and lectures (WATSON, 1963, 1964, 1965). The polysomes, their structure and function, were discussed in detail in a series of articles by A. RICH and H. NOLL (RICH, 1963; RICH et al., 1963; NOLL, 1965). The fundamental questions of genetic coding in connection with the problem of translation on the ribosome were put forward and resolved in the classic works of F. CRICK (CRICK, 1958, 1966). In the monograph we have made no effort to repeat the enumerated surveys and theoretical articles.

We shall welcome any comments and criticisms on any questions treated in the monograph.

Moscow, January 1968

A. S. SPIRIN
L. P. GAVRILOVA

Contents

Part Two

Functioning of the Ribosome

Fundamental Introduction

1. General Scheme of Protein Biosynthesis

Among the fundamental, trend-setting achievements in the quest for the understanding of the biosynthesis of proteins was the assertion that nucleic acids play a decisive rôle on this process. This hypothesis was advanced at the beginning of the 1940's. Since then, the resolution of the biosynthesis of proteins has been inseparably linked with studies on the nucleic acids and their complexes with proteins.

Our present understanding permits us to stipulate the following general scheme of this complex and multistage process (see Fig. 1).

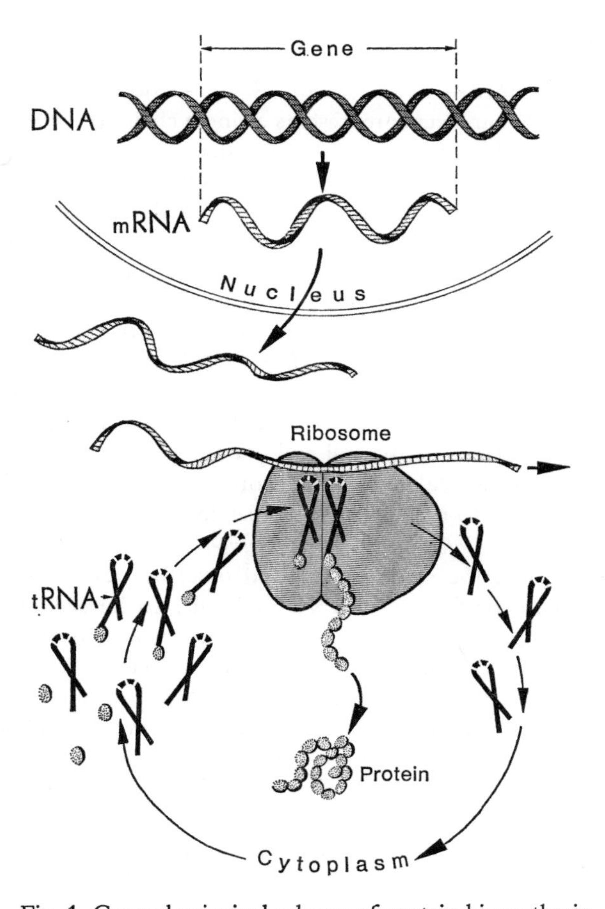

Fig. 1. General principal scheme of protein biosynthesis

The "command" rôle in the determination of the specific structure of proteins is played by the so-called *deoxyribonucleic acid* — DNA. The DNA molecule is an extremely long linear structure, consisting of two mutually coiled polymer chains. The monomers of these chains are four kinds of deoxyribonucleotides, the sequence of which along the chain is unique and specific for each DNA molecule and each region of it. Different, rather long regions of the DNA molecule are responsible for the synthesis of different proteins. Thus one DNA molecule can determine the synthesis of a large number of functionally and chemically different proteins of the cell. Only a definite region of the DNA molecule is responsible for the synthesis of each type of protein. Such a region of the DNA molecule, determining the synthesis of some one protein in the cell, is frequently denoted by the term *cistron*. The concept of the *cistron* is now considered as equivalent to the concept of the *gene*. The unique structure of the gene — a definite sequential arrangement of its nucleotides along the chain — includes all the information on the structure of a single corresponding protein.

However, DNA and its individual functional regions, which carry information on protein structure, do not participate directly in the construction of protein molecules. The first step on the way to the realization of this information, inscribed in the DNA chains, is the so-called *transcription* process. In this process, the synthesis of a chemically related polymer — *ribonucleic acid* (RNA) — occurs on the DNA chain as on a template. The RNA molecule represents a single chain, the monomers of which are four types of nucleotides, which may be considered as a slight modification of the four types of deoxyribonucleotides of DNA. The sequence of the four types of ribonucleotides in the RNA chain that is synthesized on the DNA exactly repeats the sequence of the corresponding deoxyribonucleotides of one of the two chains of DNA. In this way, the nucleotide sequence of the genes is copied in the form of RNA molecules, i.e., the information inscribed in the structure of a given gene is entirely transcribed in RNA. A large, theoretically unlimited number of such "copies" — RNA molecules — can be taken from each gene. These RNA molecules, which are multiple "copies" of the genes, and consequently carry the same information as the genes, are sent out through the cell. It is they that enter directly into a relationship with protein-synthesizing particles of the cell and play a "personal" rôle in processes of construction of protein molecules. In other words, they carry the information from the site where it is stored to the site of its realization. Correspondingly, these RNA molecules are denoted as *informational* or *messenger RNA*, abbreviated as mRNA.

Thus, the portion of the scheme discussed describes the flow of information, passing from DNA in the form of mRNA molecules to intracellular particles that synthesize the proteins. Now we shall turn to a different kind of flow — to the flow of material from which the protein should be constructed. The elementary units — monomers — of the protein molecule are amino acids, of which there are 20 different kinds. For the construction (synthesis) of a protein molecule, the free amino acids present in the cell should be brought into a suitable flow, directed to the protein-synthesizing particle, and should there be assembled into a chain in a definite unique way, dictated by the messenger RNA. Such an involvement of amino acids — the building blocks for protein synthesis — is accomplished through the addition of free amino acids to special RNA molecules, relatively small-sized. These RNA molecules, which serve for the addition of free amino acids, are not messengers, but perform a different, adaptor, function, the meaning of which will presently become evident. The amino acids are added to one of the ends of the small chains of *adaptor* or *transfer RNA* (tRNA), one amino acid per RNA molecule. Each kind of amino acid in the cell has its own specific molecules of transfer RNA, which combine only with this kind of amino acid. It is in this form, suspended on RNA, that the amino acids are delivered to the protein-synthesizing particles.

The central factor in the process of protein biosynthesis is the merging of these two intracellular flows, the flow of information and the flow of material, in the protein-synthesizing particles of the cell. These particles are called *ribosomes*. Ribosomes are ultramicroscopic biochemical "machines" of molecular dimensions, where specific proteins are assembled from the amino acid residues delivered, according to the plan included in the messenger RNA. Although only one particle is depicted in the scheme (Fig. 1), each cell contains thousands of ribosomes. The number of ribosomes determines the general rate of protein synthesis in the cell. The size (diameter) of one ribosomal particle is about 200 Å. Chemically, the ribosome is a ribonucleoprotein: it consists of a special *ribosomal RNA* or rRNA (this is the third class of RNA molecules known to us, in addition to the messenger and adaptor RNA) and molecules of structural ribosomal protein. This combination of several tens of macromolecules forms an ideally organized and reliable "machine", possessing the ability to read the information encoded in the mRNA chain, and to realize it in the form of a finished protein molecule with a specific structure. Since the essence of the process lies in the fact that the linear assembly of the 20 types of amino acids in the protein chain is unambiguously determined by the arrangement of the four types of nucleotides in the chain of a chemically entirely different polymer — a nucleic acid (mRNA), this process, which occurs in the ribosome, is now customarily denoted by the term *translation* — translation, as it were, from the four-letter alphabet of the nucleic acid chains to the 20-letter alphabet of the protein (polypeptide) chains. It is evident that all three known classes of RNA participate in the translation process — messenger RNA, which is the object of the translation, ribosomal RNA, which plays the rôle of organizer of the protein-synthesizing ribonucleoprotein particle — the ribosome, and transfer RNA's, which perform the function of the translator. The ribosome as a whole is the minimum biological particle, within which the requisite organization of all stages of the process of synthesis in space and time is realized.

After such a general sketch of the principal scheme of protein biosynthesis as a whole, it is now advisable to consider very briefly the more particular problems, each of which has been the center of enormous concentration of research labor and thought in recent years, and the solution of which has laid the foundation for modern concepts.

2. Coding of Information (The Genetic Code)

The problem of the genetic code might be cited as the first particular problem, with which we can begin our discussion, and the solution of which has proceeded at an exceptionally rapid rate in recent years and has been accompanied by important discoveries. This problem emerges from the simple fact that proteins consist of amino acids, while the nucleic acids consist of nucleotides, monomers of an entirely different chemical nature. And yet, the nucleotide sequence in the nucleic acid chain unambiguously determines the amino acid sequence in the protein chain. Moreover,

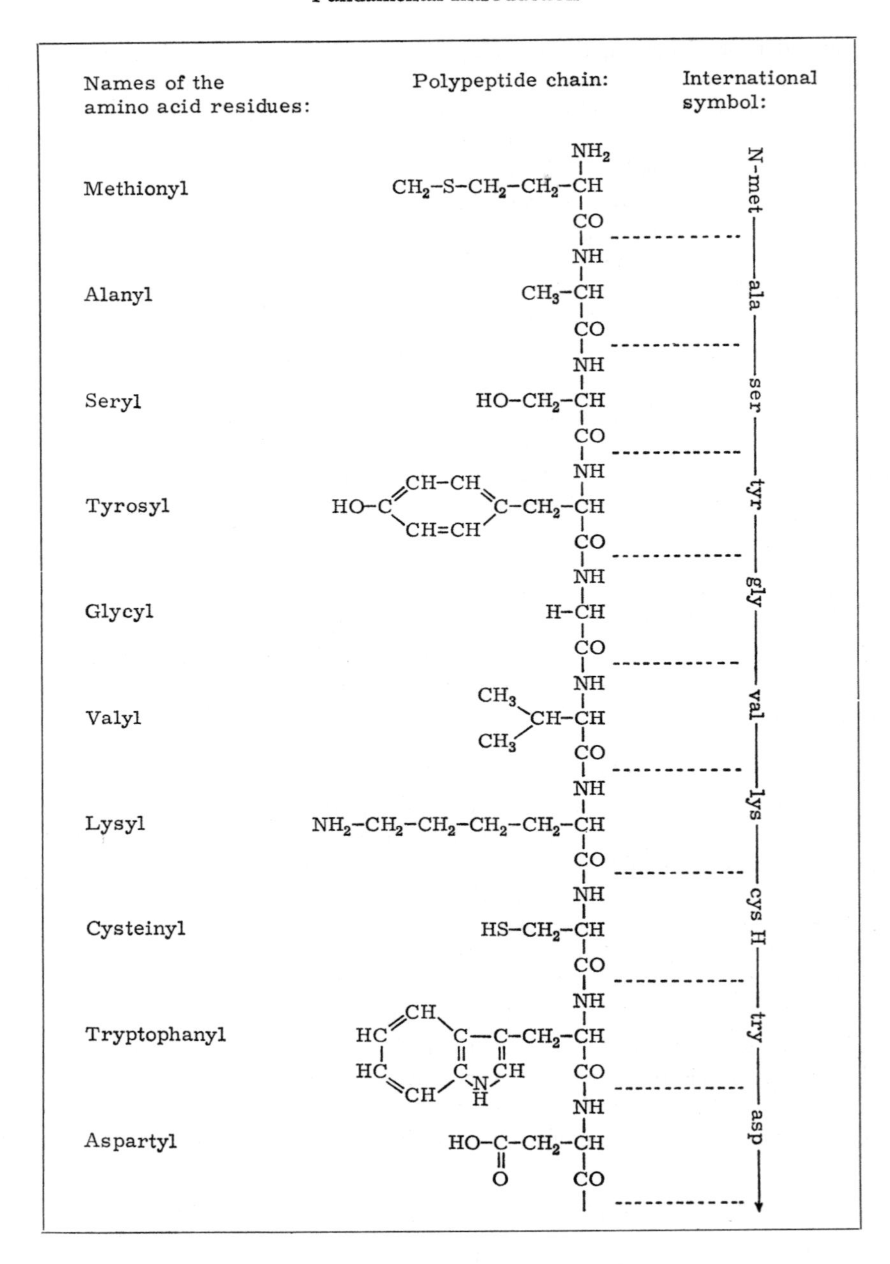

Fig. 2. Amino acid residues of natural proteins

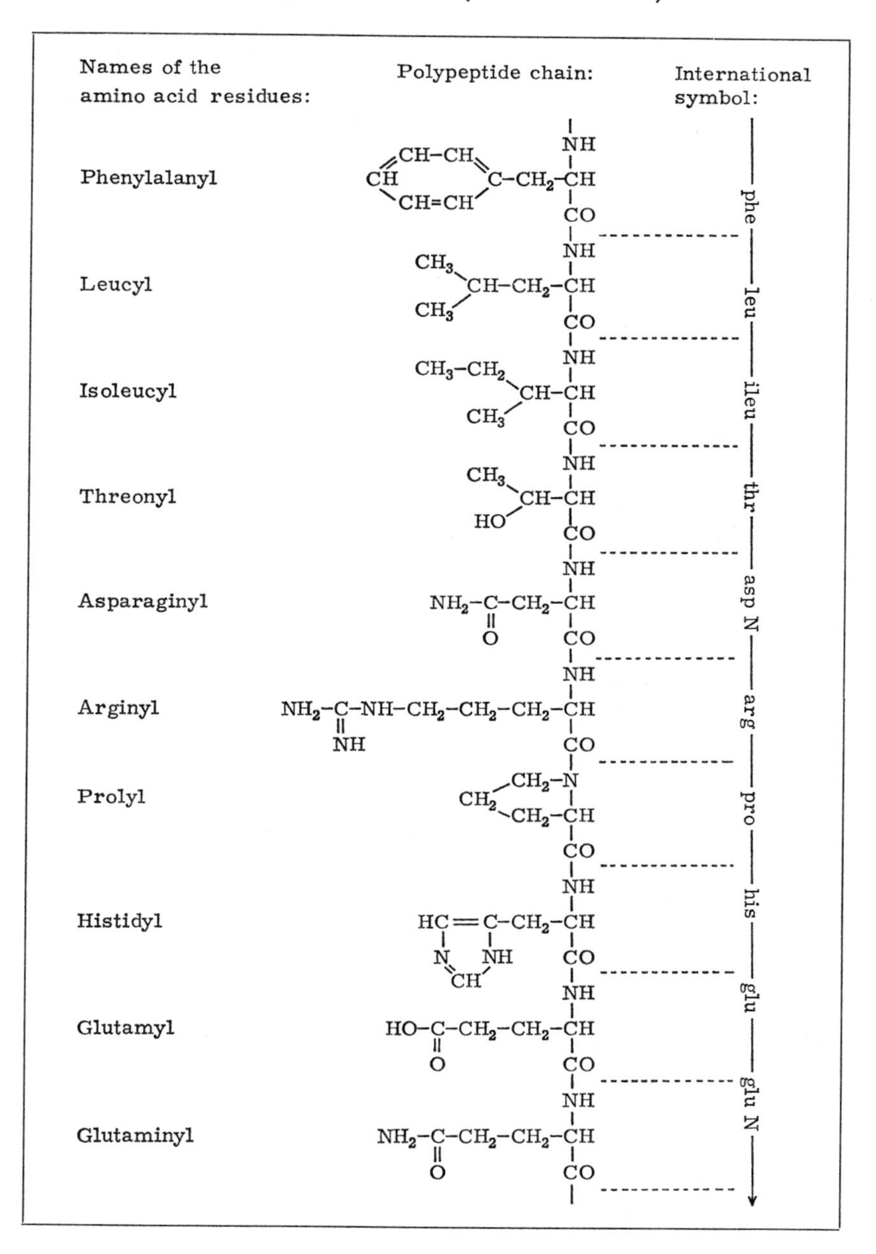

Fig. 2 (continued)

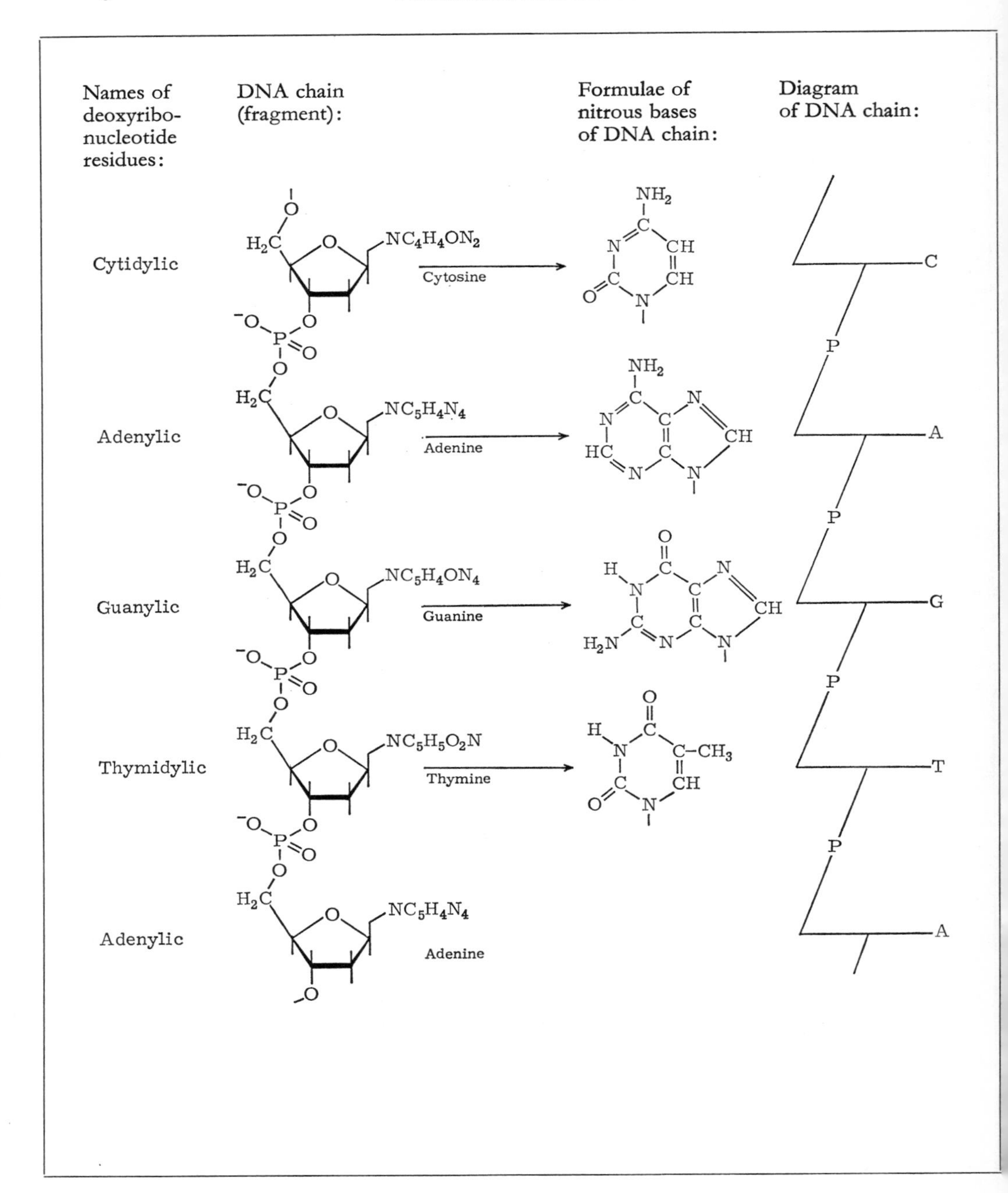

Fig. 3. DNA: polydeoxyribonucleotide chain (fragment) and formulas of its nitrogenous bases

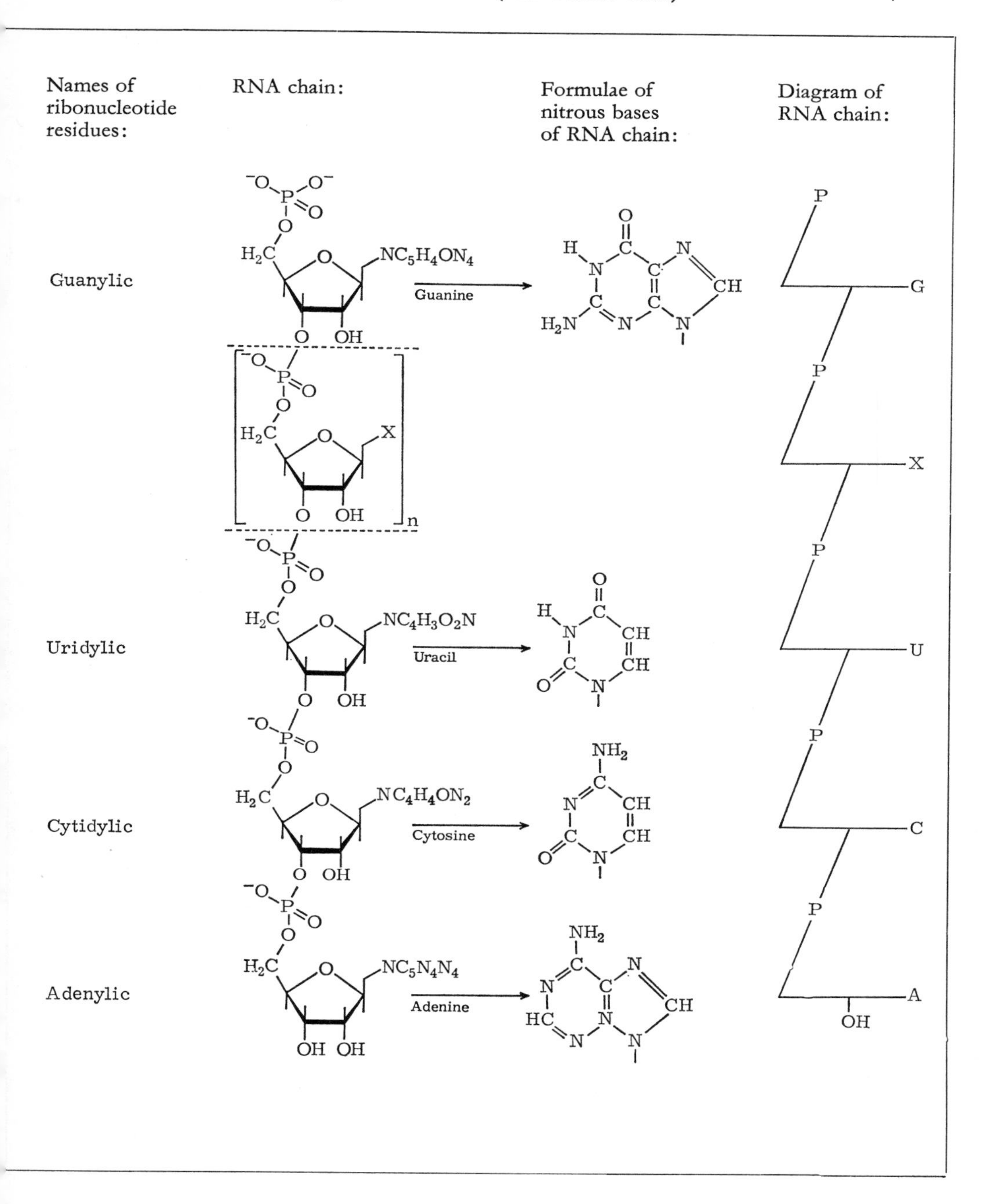

Fig. 4. RNA: polyribonucleotide chain and formulas of its nitrogenous bases

there are 20 different kinds of amino acids in a protein (Fig. 2), while there are only four different kinds of nucleotides in nucleic acids. In DNA these four kinds of nucleotides are adenylic, guanylic, cytidylic, and thymidylic deoxyribonucleotides — they can be abbreviated as A, G, C, and T, respectively (see Fig. 3). RNA is constructed from similar nucleotides — adenylic, guanylic, cytidylic, and uridylic ribonucleotides, also denoted as A, G, C, and U, respectively (Fig. 4). (Thymine, which is contained in the thymidylic nucleotide in DNA, is a methylated derivative of the uracil in RNA, and T is functionally essentially equivalent to U.) The problem consists of the following: how does the linear sequence of four kinds of elements (nucleotides) determine the linear sequence of 20 kinds of elements of an entirely different chemical nature? Back in the 50's, a hypothesis was proposed, which has now received extensive confirmation, that definite different combinations of several nucleotides correspond to different amino acids (GAMOW, 1954; GAMOW et al., 1956; CRICK et al., 1957; CRICK, 1958). In other words, each amino acid has its own definite combination of several nucleotides, which chemically have nothing in common with the amino acid. A unique code is obtained, in which each letter of the 20-letter alphabet corresponds to a definite combination of several letters of the four-letter alphabet. It has now been shown rather reliably that precisely a triplet of nucleotides (three successively bonded nucleotides) in the chain of the nucleic acid corresponds to each amino acid in the protein (CRICK et al., 1961; CRICK, 1963). The number of possible different combinations of four types of nucleotides in groups of three is equal to 64. But there are only 20 different amino acids. Thus, the number of possible triplets is quite sufficient, and there is even a large excess for the coding of 20 amino acids.

One of the greatest achievements in biology of recent years has been the experimental deciphering of the genetic code, i.e., the establishment of the concrete nucleotide composition and sequence of the triplets for all 20 amino acids contained in proteins. The first decisive experiment along this line, conducted by M. NIRENBERG and J.H. MATTHAEI (NIRENBERG, MATTHAEI, 1961), indicated that if a chain composed solely of uridylic nucleotides — polyU — is present as a messenger RNA in protein synthesis, then the synthesized polypeptide chain consists of only one kind of amino acid, namely, only of the amino acid phenylalanine. From this it has become clear that in principle the nucleotide triplet UUU codes precisely this amino acid — phenylalanine.

The further development of work on deciphering of the code has proceeded along two lines: in the first place, and chiefly, along the line of utilization of synthetic polyribonucleotides of known composition and sequence as messenger RNA in a protein-synthesizing system (SPEYER et al., 1963; NIRENBERG et al., 1963, 1965; MORGAN et al., 1966); in the second place, along the line of study of the amino acid replacements in protein as a result of mutations induced by chemical influences upon the nucleic acid (WITTMANN, WITTMANN-LIEBOLD, 1963; YANOFSKY, 1963; BRENNER et al., 1965). Now a complete code table, containing all 64 possible nucleotide triplets, can be given (Fig. 5); 61 of them are "sense codons," i.e., represent triplets that code one or another amino acid; three triplets, or "codons" (UAG, UAA, UGA) proved to be "nonsense" codons, i.e., do not code any of the known amino acids. The code, as can be seen, is highly degenerate: most of the amino acids are coded by more than one codon.

First base of codon:	Second base of codon: U		C		A		G		Third base of codon:
U	UUU UUC	Phe	UCU UCC UCA UCG	Ser	UAU UAC	Tyr	UGU UGC	Cys	U C
	UUA UUG	Leu			UAA UAG		UGA UGG	Try	A G
C	CUU CUC CUA CUG	Leu	CCU CCC CCA CCG	Pro	CAU CAC	His	CGU CGC CGA CGG	Arg	U C
					CAA CAG	GluN			A G
A	AUU AUC AUA	Ileu	ACU ACC ACA ACG	Thr	AAU AAC	AspN	AGU AGC	Ser	U C A
	AUG	Met			AAA AAG	Lys	AGA AGG	Arg	G
G	GUU GUC GUA GUG	Val	GCU GCC GCA GCG	Ala	GAU GAC	Asp	GGU GGC GGA GGG	Gly	U C
					GAA GAG	Glu			A G

Fig. 5. RNA-amino acid code

3. Storage and Replication of the Coded Information

The second problem associated with the "cybernetics" aspect of protein biosynthesis is the problem of storage and replication of the coded genetic information. From the general scheme of protein synthesis, it can be seen that the starting point for the flow of information for protein biosynthesis in the cell is DNA. Consequently, it is precisely DNA that contains the primary inscription that should be conserved and reproduced from cell to cell, from generation to generation.

Briefly touching upon the question of the site of the storage of genetic information, i.e., of the localization of DNA in the cell, the following might be stated. It has long been known that in contrast to all the other components of the protein-synthesizing apparatus, universally distributed over all portions of the living cell, DNA possesses a special, extremely limited localization: the site of its occurrence in cells of higher (eukaryotic) organisms is the cell nucleus. In lower (protokaryotic) organisms, which do not have a formed cell nucleus — bacteria and blue-green algae — DNA is also set off from the remainder of the protoplasm in the form of one or several compact nucleoid formations. Accordingly, the nucleus of Eukaryotes or the nucleoid of Protokaryotes has long been considered as the site of the genes, as a unique cellular organoid, controlling the realization of the genetic characteristics of organisms and their transmission through the generations. The genetic data on the "one-man management" of the nucleus in the cell has always been directly combined with biochemical data on the unique localization of DNA in the nucleus. In recent years, however, it has become known that such numerous cell organoids as mitochondria and plant chloroplasts contain their own DNA, thus representing to a definite degree genetically "autonomous" systems within a single cell, which on the whole is "subordinate" to the nucleus.

One of the principal questions of genetics at the molecular level is the replication of genetic information. The molecular mechanism of the exact replication of genetic information was unraveled in 1953, when J. WATSON and F. CRICK proposed a model of the macromolecular structure of DNA [WATSON, CRICK, 1953 (1, 2)]. The formulation of the basic premises lying at the basis of the model was prepared by chemical investigations of the composition of DNA, performed by E. CHARGAFF (CHARGAFF, 1950, 1951) and directly by the X-ray diffraction studies of M. WILKINS, R. FRANKLIN, and others (WILKINS et al., 1953; FRANKLIN, GOSLING, 1953). That time may be considered as the "official" birthdate of molecular biology. Indeed, the mechanism of one of the fundamental biological phenomena — the replication of self-likeness — was understood for the first time on the grounds of purely structural considerations based solely on the structure of the DNA molecule.

The main principle lying at the basis of the macromolecular structure of DNA is the so-called *principle of complementariness.* The essence of this principle is presented in Figs. 6 and 7. As has already been mentioned, the DNA molecule consists of two polynucleotide chains twisted about each other. These chains are joined together through an interaction of their opposite nucleotides. Moreover, from structural considerations, the existence of such a double-stranded structure is possible only if the opposite nucleotides of the two chains are sterically complementary, i.e., supplement one another by their three-dimensional structures. The only such mutually supplementing — complementary — pairs of nucleotides are the pairs A—T and G—C (Fig. 6). Other variations of the interaction among the four kinds of opposite nucleotides in the native DNA molecule are impossible. Consequently, according to this principle of complementariness, if we have some sequence of the four types of nucleotides in one chain of the DNA molecule, then in the second chain the nucleotide sequence will be unambiguously determined, so that T in the second chain will correspond to each A in the first chain, A in the second chain will correspond to each T in the first chain, C in the second chain will correspond to each G in the first chain, and G in the second chain will correspond to each C in the first chain (Figs. 6 and 7).

It is evident that the indicated structural principle, lying at the basis of the double-stranded structure of the DNA molecule, permits an easy understanding of the exact reduplication of the initial structure, i.e., the exact reduplication of the information inscribed in the chains of the molecule in the form of a definite sequence of the four kinds of nucleotides. Actually, the synthesis of new DNA molecules in the cell occurs only on the basis of the DNA molecules already present. In this case, the two chains of the initial DNA molecule begin to separate at one of the ends, and a second chain begins to be assembled on each of the separated single-stranded sections from the free nucleotides present in the medium, in exact correspondance to the principle of complementariness (Fig. 7). The process of separation of the two chains of the initial DNA molecule continues, and correspondingly, the two chains are supplemented by complementary chains. As a consequence, as can be seen from the scheme, two DNA molecules, exactly identical with the original molecule, appear in place of one. In each of the resulting "daughter" DNA molecules obtained, one chain is entirely from the original molecule, while the other is newly synthesized (MESELSON, STAHL, 1958). The main thing that still must be emphasized is that the potential ability for exact reduplication is lodged in the double-stranded complementary structure of DNA as such, and the discovery of this is unquestionably one of the main achievements of modern biology.

However, the problem of reduplication — replication — of DNA is not exhausted by an ascertainment of the *potential ability* of its structure for exact reproduction of its nucleotide sequence. Actually, DNA itself is not at all a self-reproducing molecule. The accomplishment of the process of synthesis — reduplication — of DNA according

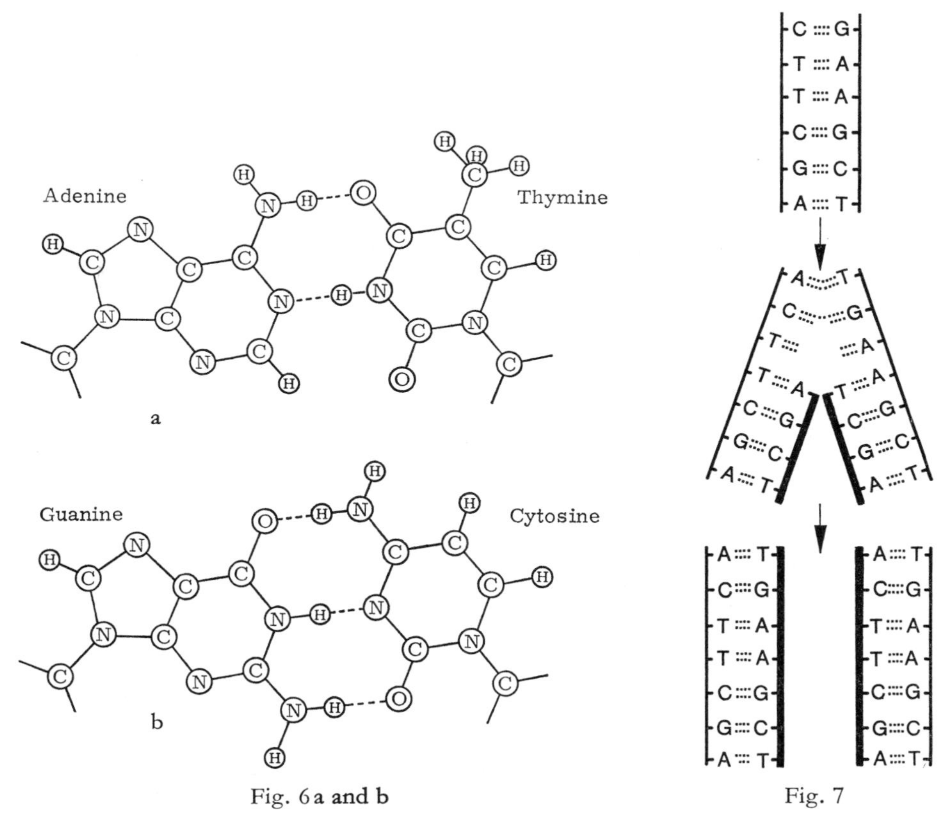

Fig. 6a and b. Pairing of nitrogenous bases in DNA: A – T and G – C pairs

Fig. 7. The principle of complementariness and scheme of DNA reduplication

to the scheme described above necessitates the activity of a special enzymatic protein, which is called DNA polymerase (Bessman et al., 1958; Lehmann et al., 1958; Bollum, 1963). Evidently it is precisely the enzyme that accomplishes the process of separation of the two chains which occurs successively from one end of the DNA molecule to the other, with simultaneous polymerization of free nucleotides on them according to the complementariness principle. Thus, DNA, like a template, only sets the nucleotide sequence in the chains that are synthesized, while the process itself is conducted by a protein. The work of the enzymatic protein during DNA replication is one of the most interesting problems of the present day. Evidently the protein seems to slide actively along the double-stranded DNA molecule from one end of it to the other, leaving a doubled "reduplicated" tail behind it. The physical principles of the work of this protein are still obscure.

4. Transfer of Information (Transcription)

The next problem, determining one of the most important aspects of protein biosynthesis, is the problem of the transfer of information from DNA to the protein-synthesizing particles — the ribosomes. The first and basic point in this transfer of information is *transcription* — the synthesis of messenger RNA on regions of the DNA chains (on genes).

The idea itself that DNA is not an immediate participant in the process of protein biosynthesis, but carries out its "directing" function in one way or another through

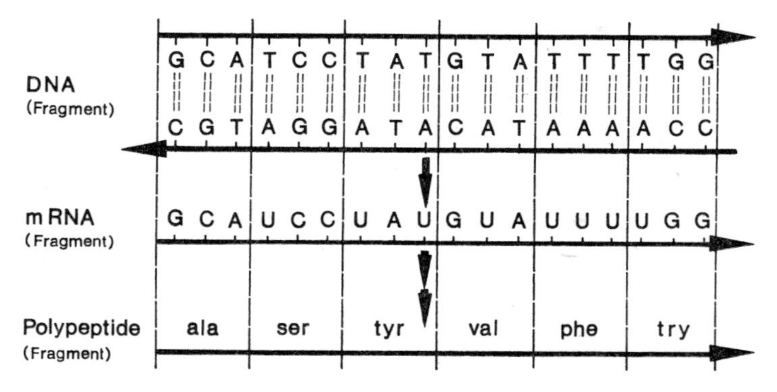

Fig. 8. Colinear relationships of the monomer sequences of DNA, RNA, and protein chains

the agency of RNA, arose comparatively long ago, evidently having been first proposed in the 1940's by T. Caspersson (Caspersson, 1941). Thus, the famous hierarchy "DNA→RNA→protein" was born. Subsequently it was determined that the mediator, "copying" DNA, is by no means all the cellular RNA, but only a special fraction of it, comprising a small portion of the total RNA [Spirin et al., 1957; Belozersky, Spirin, 1958, 1960; Brenner et al., 1961; Gros et al., 1961 (1, 2); Spiegelman, 1961; Jacob, Monod, 1961]. This fraction has been named the informational or messenger RNA (mRNA).

It has been found that the messenger RNA chain is synthesized, directly utilizing the corresponding section of DNA as a template (Weiss, Nakamoto, 1961; Geiduschek et al., 1961; Hurwitz et al., 1962; Hurwitz, August, 1963). The mRNA chain synthesized in this case copies the nucleotide sequence of one of the two chains of DNA (assuming that U in RNA corresponds to its derivative T in DNA). This occurs on the basis of the same structural principle of complementariness which determines DNA reduplication (Fig. 8). It has been found that when mRNA is synthesized on DNA in the cell, only one chain of DNA is used as the template for the formation of the mRNA chain (Marmur et al., 1963). To each G of this DNA chain a C in the RNA chain being constructed will then correspond, to each C of the DNA chain a G in the RNA chain, to each T of the DNA chain an A in the RNA chain, and to each A in the DNA chain a U in the RNA chain. As a result, the RNA chain obtained will be strictly complementary to the template DNA chain and consequently identical in nucleotide sequence (assuming T = U) with the second chain of DNA. Thus, there is a "rewriting" of the information from DNA to RNA, i.e., a transcription. The transcribed nucleotide combinations of the RNA chain directly determine the assembly of the corresponding amino acids coded by them into a protein chain (Fig. 8).

And again, just as in the discussion of DNA reduplication, its enzymatic character must be indicated as one of the most important points of the process of transcription. DNA, which is the template in this process, entirely determines the nucleotide sequence in the mRNA chain being synthesized and the entire specificity of the RNA formed, but the course of the process itself is accomplished by a special enzymatic protein (HURWITZ et al., 1960, 1961; WEISS, 1960; STEVENS, 1960). This enzyme is called the RNA polymerase. Its molecule possesses a complex organization, which permits it to move actively along the DNA molecule, simultaneously synthesizing an RNA chain complementary to one of the DNA chains. The DNA molecule serving as the template is not expended in this case and is not altered irreversibly, remaining in its former state and always ready for such a transcription of an unlimited number of "copies" — mRNA — from it. The flow of these mRNA molecules from DNA to the ribosomes represents the flow of information which provides the entire programming of the protein-synthesizing apparatus of the cell, the entire aggregate of its ribosomes.

5. Involvement of Amino Acids in Protein Synthesis

Among the principal problems of protein biosynthesis in the living cell being considered, a special place is occupied by problems associated not with the "genetic or "programmed" provision of protein synthesis, but with the involvement and flow of the material and energy itself for the construction of proteins. Obviously, this "material" aspect of protein synthesis is no less important and decisive than its "cybernetic" aspect. It is precisely at the level of activation and involvement of amino acids in the process of protein synthesis that the two principal problems of biosynthesis are resolved: the energy provision of the process and the primary "recognition" by the amino acids of the nucleotide combinations corresponding to them.

Both these problems are solved through the formation of compounds of the amino acids with molecules of transfer RNA, or tRNA (HOAGLAND et al., 1957, 1958; OGATA et al., 1957; HOAGLAND, 1960). At first there is an energetic "activation" of the amino acid on account of its enzymatic reaction with a molecule of adenosine triphosphate (ATP), and then the "activated" amino acyl combines with the end of the relatively short tRNA chain; the increase in the chemical energy content of the activated amino acid is conserved in this case in the form of the energy of the chemical bond between the amino acid and tRNA. The amount of this free energy is sufficient for the subsequent formation of a peptide bond between amino acids during the construction of protein on the ribosome. In this way the energy aspect of the process of polymerization of amino acids into a protein chain is provided for.

But the second problem is also solved simultaneously. Actually, the reaction between the amino acid and a tRNA molecule is brought about by an enzyme denoted as aminoacyl-tRNA synthetase. Each of the 20 types of amino acids has its own special enzymes, which bring about the reaction with the participation only of a certain type of amino acid. Thus, there are no less than 20 groups of enzymes (aminoacyl-tRNA synthetases), each of which is specific for one type of amino acid. On the other hand, each of these enzymes can bring about a reaction not with any tRNA molecule, but only with those that carry a strictly defined combination of nucleotides in their chain. For each enzyme, these combinations of nucleotides in the tRNA chain are different. Thus, we find that, for example, the enzyme phenylalanyl-tRNA syn-

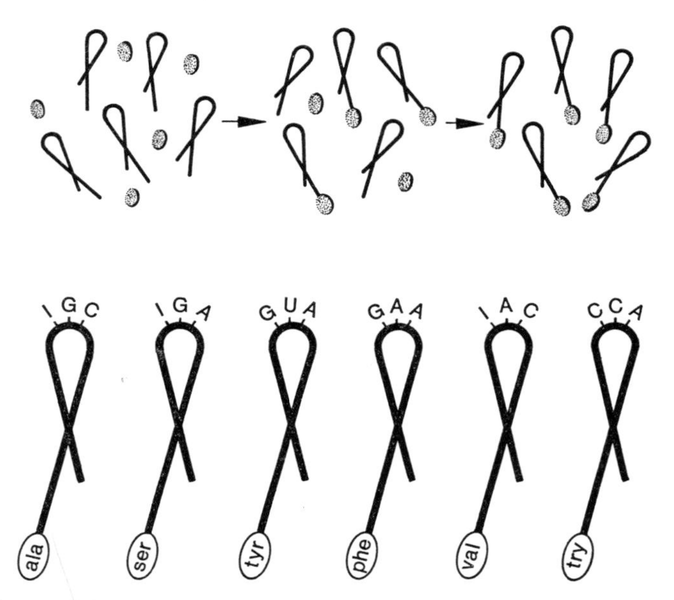

Fig. 9. Attachment of amino acid residues to transfer RNA (tRNA) molecules. *Lower* — specific ascription of amino acids to definite tRNA's. (I of tRNA is inosinic nucleotide, deaminated derivative of A.)

thetase brings about a reaction only between the amino acid phenylalanine and a special tRNA, which, in particular, carries the combination of nucleotides GAA in the middle of the chain; therefore the enzyme can add phenylalanine only to this and to no other tRNA. In exactly the same way, the enzyme tryptophanyl-tRNA synthetase brings about a reaction only with the participation of the amino acid tryptophan and another specific tRNA, which, in particular, evidently has the combination CCA in the middle of the chain; correspondingly, only a compound of tryptophan with this tRNA will be formed in the enzymatic reaction. These unambiguous relationships, due to the substrate specificity of the corresponding enzymatic proteins, are schematically represented in Fig. 9. Thus, as a result of the existence of a set of such specific enzymes, recognizing, on the one hand, the nature of the amino acid, and on the other hand, the nucleotide sequence of tRNA, each of the 20 kinds of amino acids is proved to be ascribed only to a definite tRNA with a given characteristic nucleotide combination. (This does not at all mean, however, that the enzyme "recognizes" tRNA precisely according to the given nucleotide combination in the middle of the chain; more likely the enzyme recognizes its own tRNA according to some different regions of the nucleotide sequence, although the recognizable regions are somehow correlated with the presence of the indicated triplets.)

6. Synthesis of Protein on the Ribosome (Translation)

It is precisely as a result of such specific "charging" of amino acids on the tRNA molecules before they are delivered to the ribosome that makes it possible for the

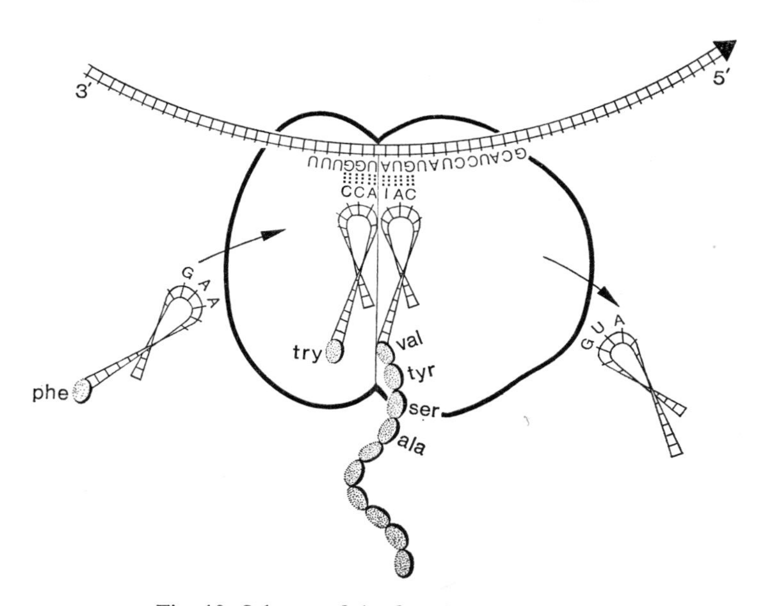

Fig. 10. Scheme of the functioning ribosome

tRNA molecules to perform their adaptor functions in the interaction with the triplets of messenger RNA during the process of translation on the ribosome. Here we come to the central problem of protein synthesis — the mechanism of synthesis of the protein chain with the participation of the ribosome, messenger RNA, and adaptor RNA's charged with amino acids (aminoacyl-tRNA). Certain points of this process, insofar as we envision them at the present day, are given schematically in Fig. 10.

In Fig. 10, it is evident first of all that the messenger RNA molecule is associated with the ribosome, or, as sometimes said, the ribosome is "programmed" with messenger RNA. At each given moment, only a relatively short section of the mRNA chain is on the ribosome itself. But it is precisely this section that can interact with the transfer RNA molecules with the participation of the ribosome. And here again the main role is played by the principle of complementariness, already analyzed twice above. Thus, if the section of the mRNA chain situated in the corresponding site of the ribosome possesses the nucleotide sequence UGG (read in the direction from the starting end of the mRNA chain to the terminal end),[1] then a chain of tRNA carrying a complementary triplet CCA in its "contact" region (anticodon) is automatically attached to this triplet. But since, as has already been indicated above, a tRNA molecule with such a triplet CCA may be charged only with the amino acid tryptophan, consequently the triplet UGG of messenger RNA determines the delivery into the ribosome and binding according to the complementariness principle precisely of tryptophanyl-tRNA. Here lies the explanation of the mechanism of why a given triplet of the mRNA chain corresponds to its own definite amino acid. Evidently the necessary intermediate link, or adaptor, in the "recognition" of its own triplet on mRNA by each amino acid is the adaptor RNA (tRNA)[2].

Furthermore, it is evident from the scheme (Fig. 10) that on the ribosome, in addition to the tRNA molecule charged with amino acid just considered, there is another tRNA molecule. But in contrast to the tRNA molecule discussed above, the end of this tRNA molecule is connected to the end of the growing protein (polypeptide) chain. This situation reflects the dynamics of the events occurring on the ribosome during the process of synthesis of a protein molecule. These dynamics may be represented in the following way. Let us begin with some intermediate moment, shown in the scheme and characterized by the presence of a protein chain that has already begun to be constructed, with tRNA bonded to it, and a new tRNA molecule charged with the corresponding amino acid which has just entered the ribosome and is attached to the triplet. Evidently the very act of attachment of a tRNA molecule to the mRNA triplet situated at a given site on the ribosome leads to such a mutual orientation and close contact between the amino acid residue and the protein chain under construction that a covalent (peptide) bond arises between them. The bond arises in such a way that the end of the protein chain under construction, shown in the scheme connected to tRNA, combines with the amino acid residue of the arriving aminoacyl-tRNA. As a result, the "right-hand" tRNA is free, while the protein chain is lengthened by one amino acid and is connected to the "left-hand" (just arrived) tRNA. After this, the "left-hand" tRNA is thrown over "to the right," together with the mRNA codon bound to it; then the previous molecule of tRNA that is free now is displaced and leaves the ribosome; a new tRNA charged with the protein chain under construction, lengthened by one amino acid residue, takes its place; and the mRNA chain will be shifted relative to the ribosome by one triplet to the right. As a result of the shift of the mRNA chain by one triplet to the right, the following vacant triplet appears on the ribosome (on the scheme — UUU), and the corresponding tRNA charged with an amino acid (phenylalanyl-tRNA) is immediately attached to it

[1] Any polynucleotide chain is polar. Therefore one of its ends may be denoted as the starting end of the chain and the other as the terminal end of the chain. The starting end of the chain is taken to be the so-called 5′ end of the polynucleotide; the other end of the polynucleotide, the 3′ end, is considered as the terminal (see Part II, Section III).

[2] It is necessary to note that in the interaction of the mRNA triplet with tRNA, complementariness is not so strict as in the cases of reduplication and transcription. Thus, here, besides the canonic pairs G – C and A – U, there is a possibility, with the triplet occupying one of the extreme positions, of the formation of the pairs G – U, I – U, I – C and I – A (where I is the inosinic nucleotide representing the deaminated derivative of A and is met in tRNA) (CRICK, 1966). That is why the triplet GCA on mRNA binds the combination IGC of alanyl-tRNA, the triplet UCC binds IGA of seryl-tRNA, the triplet UAU binds GUA of tyrosyl-tRNA, the triplet GUA binds IAC of valyl-tRNA, and the triplet UUU binds GAA of phenylalanyl-tRNA (see Figs. 9 and 10).

according to the complementariness principle. This again causes the formation of a covalent (peptide) bond between the protein chain under construction and the phenylalanine residue, and immediately thereafter a shift of the mRNA chain by one triplet to the right, with all the consequent results. And so forth. In this way, a sequential, triplet after triplet, passing of the messenger RNA chain through the ribosome is accomplished, as a result of which the mRNA chain is "read" by the ribosome completely, from beginning to end. At the same time, and coupled with this, there is a sequential, amino acid after amino acid, buildup of the protein chain. Correspondingly, tRNA molecules charged with amino acids enter the ribosome, one after another, and tRNA molecules without amino acids leave. In the solution outside the ribosome, the free tRNA molecules again combine with amino acids and again carry them into the ribosome, themselves returning in a cyclic manner, without breakdown or change.

Of course, the most striking and mysterious moment in the dynamic model of translation described, and perhaps in the entire problem of protein biosynthesis, is the extremely coordinated, reliable, and multifaceted work of the small particle called the ribosome and consisting only of several dozen macromolecules. This forces us to think very seriously of unknown structural principles, lying at the basis of the organization of the ribosome, this astonishing molecular "machine".

Part One

Structure of the Ribosome

At the very beginning of the exposition of the problem, we should obviously state that the structure of the ribosomes is not yet known. Therefore, everything that will be outlined here pertains chiefly to the description: a) of the general physical properties and chemical composition of the ribosomes; b) of the structure of the components of the ribosomes, RNA and protein; c) of different types of structural transformations of the ribosomes, including disassembly and reconstitution. Of course, the experimental development and analysis of these problems is the way leading to an understanding of the structure of the ribosomes. However, this path has not yet led us to the point of being able to give an integral description of the construction of the ribosomal particle on the basis of the indicated scattered groups of data, not to mention a description of the mode of its functioning.

I. Physical Properties and Chemical Composition of the Ribosomes

1. Shape and Size

On ultrathin sections of tissues and cells, the ribosomes are visible under an electron microscope as dense, roundish, granules, 150 to 250 Å in diameter.

An electron microscopic study of isolated ribosome preparations permits a more accurate determination of their shape and size. In a first approximation, the ribosome may be described as a *slightly prolate ellipsoid of revolution* with an axial ratio no greater than 1.5. Actually, the detailed shape of the ribosome is more complex, specific, and cannot be described by a simple geometrical figure. The shape of the ribosomes varies very little from one biological object to another.

In dimensions and molecular weight, all the ribosomes studied thus far break down into two distinct groups, depending upon their biological source. One group is formed by the relatively small ribosomes of bacteria and blue-green algae. Another group is comprised of the somewhat larger ribosomes of all the other organisms that possess an organized cell nucleus, i.e., the ribosomes of all the eukaryotic organisms, including animals, higher plants, algae and fungi. The somewhat larger size of eukaryotic ribosomes in comparison with bacterial ribosomes is due to two factors: 1. their absolute content of RNA is approximately 30 to 40% greater than in the bacterial ribosomes; 2. they are more loaded with protein than the bacterial ribosomes, i.e., possess a greater relative protein content (see below, Sections I, 4; II and III).

a) 70 S Ribosomes of Protokaryotes (Bacteria and Blue-green Algae)

One of the most widely used characteristics of the ribosomes is the sedimentation coefficient, which is directly correlated with their size and molecular weight (see, for example, INOUYE et al., 1963).

A special comparative examination of ribosome preparations from 25 species of bacteria, two species of blue-green algae, and 26 species of fungi (TAYLOR, STORCK, 1964) indicated that among all the Protokaryotes studied (bacteria and blue-green algae), the sedimentation coefficient of the ribosomes, $S^0_{20,w}$ is about *70 S* (average value 69 S, total range of variation from 64 to 72 S in different species). A typical and most studied representative, of course, is the bacterium *Escherichia coli*, the ribosomes of which, as was well shown in the classic investigations of TISSIÈRES, WATSON et al. (TISSIÈRES, WATSON, 1958; TISSIÈRES et al., 1959), possess a sedimentation coefficient of 69.1 S and a molecular weight of 2.8×10^6 ($2.6 \times 10^6 \pm 6\%$ from S and D, $3.1 \times 10^6 \pm 18\%$ from S and $[\eta]$). According to the data of electron microscopy, the dimensions of the dry particle are $170 \times 170 \times 200$ Å (HALL, SLAYTER, 1959; HUXLEY, ZUBAY, 1960); the approximation by a slightly prolate ellipsoid of revolution gives a "dry" molecular volume of about 3×10^6 Å^3. However, evidently when the particles are allowed to dry out on a substrate, they may be compressed to some degree or another, so that the dimensions of the 70 S ribosome in aqueous medium may be greater. On the basis of data on the electron microscopy of frozen-dried ribosomal particles (HART, 1962, 1963) and data on the scattering of X-rays in concentrated gels of ribosomal particles (LANGRIDGE, HOLMES, 1962), it may be assumed that in aqueous medium the 70 S ribosome of *E. coli* possesses a length (large axis of the ellipsoid of revolution) of up to 290 Å and a width (small axis of the ellipsoid) of about 210 Å. This corresponds to a volume of the particle in solution of up to 7×10^6 Å^3.

The similarity of the sedimentation coefficients of the ribosomes from all the species of bacteria and blue-green algae studied permits us to think that they are all characterized by approximately the same shape and size as the ribosome of *E. coli*. Thus, the molecular weight of the 70 S ribosomes of protokaryotic organisms is about 3×10^6, and they can be approximated by prolate ellipsoids of revolution with dimensions of $200 \times 170 \times 170$ Å in the dry state and up to $290 \times 210 \times 210$ Å in solution.

More detailed physical characteristics of the ribosomes of *E. coli* are enumerated in Table 1.

b) 80 S Ribosomes of Eukaryotes

The sedimentation coefficients of the ribosomes from eukaryotic organisms are substantially higher than 70 S. For example, the average value of the sedimentation coefficient $S^0_{20,w}$ for the ribosomes from 26 species of fungi was found to be equal to 81 S, with a range of variation from 79 to 85 S (TAYLOR, STORCK, 1964). According to the most reliable data, the sedimentation coefficients of purified ribosomes from mammalian tissues are also about 80 S. Thus, for rat liver ribosomes and rat liver tumor (Jensen sarcoma) ribosomes, a careful determination of the sedimentation coefficient gives a value of 83 S (HAMILTON et al., 1962; PETERMANN, 1960); for ribosomes from the Novikoff rat hepatoma, other authors obtained values of the sedimentation coefficient of 79 to 81 S (KUFF, ZEIGEL, 1960); a value of 77 S was obtained from guinea pig liver ribosomes (TASHIRO, SIEKEVITS, 1965); rabbit reticulocyte

Table 1. *Physical properties of the ribosomes*

Properties	70 S ribosomes of *E. coli*[a]	80 S ribosomes of Eukaryotes[b]
Sedimentation coefficient, $S^0_{20,w}$, svedbergs	69.1	81 ± 4
Intrinsic viscosity, $[\eta]$, cm³/g	6.1	4.5 – 8
Diffusion coefficient, $D^0_{20,w} \times 10^7$ cm²/sec	1.83	1.3 – 1.4
Specific partial volume, $\bar{v}$, cm³/g	0.64	0.66 – 0.67
Molecular weight		
From S and D	2.6×10^6	$4.1 - 4.7 \times 10^6$
From S and $[\eta]$	3.1×10^6	$4.1 - 4.5 \times 10^6$
From sedimentation equilibrium	—	$4.8 - 5.2 \times 10^6$
From dry electron microscopic volume	3.4×10^6	$4.1 - 4.6 \times 10^6$
Dimensions in dried state, from EM:		
Linear dimensions, Å	200 × 170 × 170	240 × 200 × 200
Volume, Å³	3×10^6	5×10^6
Dimensions in aqueous medium, calculated approximately from EM of frozen-dried particles, with corrections for flattening:		
Linear dimensions, Å	290 × 210 × 210	340 × 240 × 240
Volume, Å³	7×10^6	10×10^6
Probable amount of retainable water, w, g/g	0.9	0.7

[a] According to the data of Tissières et al., 1959, and Hall, Slayter, 1959; the dimensions in aqueous medium and probable w were derived from indirect data — see text.

[b] According to literature data, cited in the text. Dimensions in aqueous medium and probable w were derived from indirect data.

ribosomes were reported to possess a sedimentation coefficient of about 80 S (Dintzis et al., 1958). Unfortunately, no one has as yet made a broad comparative analysis of the ribosomes of different animal tissues, so that at the present we must be content with extremely scattered data, obtained at different times and in different laboratories. As a result, of course, the apparent variability of the values increases. The lowest values of the sedimentation coefficient have been reported for ribosomes from the calf pancreas — 75 S (Keller et al., 1963), and for ribosomes from the calf thymus — 74 S (Hess, Lagg, 1963). However, it is not known to what extent this reflects the real range of variation, and not the dispersion of data as a result of experimental errors. In any case, other researchers have obtained higher values for ribosomes from the calf pancreas and nuclei from the calf thymus — 80 S (Madison, Dickman, 1953) and 78 S (Pogo et al., 1962), respectively. In general, for all mammalian tissues we can assume an average value of the sedimentation coefficient of their ribosomes of about 80 S ± 3 S.

The same values of the sedimentation coefficients have been found to characterize ribosomes from the cytoplasm of the cells of higher plants: Ts'o, Bonner, and Vinograd (Ts'o et al., 1956, 1958) demonstrated that ribosomes from pea seedlings possess a sedimentation coefficient of 80 S ± 1 S; later similar values of the sedimentation coefficients (80 to 83 S) were obtained for the cytoplasmic ribosomes of cabbage, clover, ryegrass, tobacco, spinach, and kidney beans (Lyttleton, 1960, 1962; Clark et al., 1964; Spencer, 1965; Boardman et al., 1965; Odintsova et al., 1967).

The higher sedimentation coefficients of the ribosomes from eukaryotic organisms in comparison with bacterial ribosomes agree with their somewhat higher molecular weight values and dimensions. The values of the molecular weights of the 80 S ribosomes, calculated by various methods (from sedimentation and intrinsic viscosity, from sedimentation and diffusion, from data on sedimentation equilibrium, from electron microscopic measurements of the dimensions of dry particles), almost always lie within the range from 4 to 5 million (CHAO, SCHACHMAN, 1956; Ts'o et al., 1956, 1958; DINTZIS et al., 1958; KUFF, ZEIGEL, 1960; TASHIRO and YPHANTIS, 1965); the lowest value — 3.6 million — was obtained from data on light scattering (HAMILTON et al., 1962).

The maximum linear dimensions of particles visible on electron micrographs of dry shadowed 80 S ribosomes from animals and higher plants are 200 to 240 Å, with a height of the particles of about 180 Å for ribosomes dried in the usual manner (HALL, cited according to DIBBLE, DINTZIS, 1960; KUFF, ZEIGEL, 1960) and up to 350 Å with the same height of 160 to 180 Å for frozen-dried particles (Ts'o et al., 1958). Negatively stained 80 S ribosomes from mammalian cells possess a length of about 280 to 300 Å and a width of about 220 Å on electron micrographs (SHELTON, KUFF, 1966). In the case of negative staining of the 80 S ribosomes from higher plants, the measured length of the particles was found to be equal to 260 ± 10 Å, and the width 190 ± 10 Å to 200 ± 10 Å (ODINTSOVA et al., 1967). On the basis of the enumerated electron microscopic data, and taking into consideration the inevitable flattening of the ribosomes during application onto the supporting surface and decrease in the volume during drying, especially in the absence of a contrasting substance, it may be considered that the 80 S ribosomes of higher organisms, being approximated by a slightly prolate ellipsoid of revolution, possess a length of about 240 Å and a thickness of about 200 Å in the dry state, and a length (large axis of the ellipsoid) of about 300 Å to 340 Å and a width (small axes) of about 200 to 240 Å in aqueous medium. The latter values correspond to a particle volume in solution of about 10×10^6 Å^3.

Thus, the 80 S ribosomes of eukaryotic organisms can be characterized as slightly prolate ellipsoids of revolution with a molecular weight of about $4-5 \times 10^6$ and approximate dimensions of $240 \times 200 \times 200$ Å in the dry state and $340 \times 240 \times 240$ Å in aqueous medium.

c) 70 S Ribosomes of Chloroplasts and Mitochondria

In all higher, eukaryotic, organisms, the ribosomes are contained not only among the basic mass of the cytoplasm and nucleus, but also in the "energy-giving" organoids of the cell, the chloroplasts and mitochondria.

It has been shown that the ribosomes of the chloroplasts of green plants differ distinctly from the cytoplasmic 80 S ribosomes of the same cells: they are somewhat smaller and are of the 70 S type, in which they exhibit a surprising similarity to the bacterial ribosomes (LYTTLETON, 1960, 1962; CLARK et al., 1964; SPENCER, 1965; BOARDMAN et al., 1965, 1966; SVETAILO et al., 1966, 1967; ODINTSOVA et al., 1967; SAGER, 1967). The exact sedimentation coefficient $S^0_{20,w}$ of the ribosomes from tobacco and pea chloroplasts is 69.8 ± 0.1 S (BOARDMAN et al., 1966; SVETAILO et al., 1966, 1967). The dimensions of the 70 S ribosomes, isolated from the chloroplasts of higher plants, were determined by electron microscopy of negatively stained samples and proved equal to 220 ± 10 Å $\times$ 170 ± 10 Å (ODINTSOVA et al., 1967).

The mitochondrial ribosomes of eukaryotic cells also were shown to be not of the 80 S type, like all the cytoplasmic ribosomes, but of the 70 S type (KÜNTZEL, NOLL, 1967).

2. Compactness

Repeated attempts have been made to evaluate the compactness of the ribosomes in solution on the basis: a) of the values of the intrinsic viscosity of ribosome preparations; b) of data on X-ray diffraction in moist gels or concentrated suspensions of ribosomes; c) of electron microscopy of frozen-dried particles, in comparison with air-dried particles.

Since the shape of the ribosome does not differ strongly from a sphere, it is at first tempting to attempt to determine the degree of compactness and swelling of the ribosomal particle on the basis of a determination of the intrinsic viscosity. Actually, it is known that the specific viscosity of ideal nonhydrated spherical particles does not depend upon their size and is equal to the volume fraction of these particles, multiplied by 2.5. If the viscosity is greater, this may be due either to a deviation from the spherical shape or to immobilization of some amount of the solvent (due to hydration, porosity, etc.). The measured intrinsic viscosity of the ribosomes naturally always proves to be appreciably greater than should have been expected for ideal nonhydrated spheres. Since many researchers have found this deviation to be rather large, i.e., the intrinsic viscosity was anomalously high, the opinion has arisen that the ribosome is an extremely hydrated and "porous" particle. For example, a value of the intrinsic viscosity $[\eta]=11$ cm^3/g has been found for ribosome preparations from pea seedlings (TS'O et al., 1958), rat liver (HAMILTON et al., 1962), and rabbit reticulocytes (TS'O, VINOGRAD, 1961); the specific partial volume of all these ribosomes is $\bar{v}=0.66-0.67$ cm^3/g; consequently, if the shape of the ribosomes is spherical, then $[\eta]=2.5\times(\bar{v}+w)$ or $w=\frac{[\eta]}{2.5}-\bar{v}=3.7$ g/g (where w is the number of grams of water per gram of ribosomes). We find that the ribosome in solution represents a highly swollen ("porous") particle, containing 80% (by weight) water. The diameter of such particles in solution should comprise about 400 Å, the volume up to 35×10^6 Å^3.

The conclusion that the structure of the ribosomal particles is highly loose has also been confirmed in a small-angle X-ray scattering study of concentrated suspensions of ribosomes from rabbit reticulocytes (DIBBLE, DINTZIS, 1960; DIBBLE, 1964). A value of 20.5×10^6 Å^3 has been calculated for the volume of the particle in solution. Since the dry volume of the 80 S ribosome was assumed equal to about 4×10^6 Å^3, the swelling of the particle was estimated as $w=2.7$ g/g.

However, all this scarcely corresponds to the actual situation. First of all, a measurement of the specific viscosity of the ribosomes is fraught with great dangers: the presence of the slightest viscous impurities in the preparations, aggregation of the particles, etc., easily introduces a large relative error into the low values of the specific viscosity of the ribosomes. The presence of aggregates and impurities presents a danger of great errors in measurements of light scattering and X-ray diffraction as well. Evidently in the case of calculations based upon the values of the intrinsic viscosity of the ribosomes, an effort should be made to take into consideration only the *lowest* of the values obtained experimentally. Thus, a lower value than that cited above was obtained for the ribosomes of rabbit reticulocytes by DINTZIS et al. (DINTZIS et al., 1958) — $[\eta]=8$ cm^3/g; from this, the calculated value for the amount of water retainable in the ribosome is lower, when an ideally spherical shape is assumed. But even this value is evidently too high, since even lower values of the intrinsic viscosity for 80 S ribosomes have been published in the literature — for example, the value 5 cm^3/g for yeast ribosomes (CHAO, SCHACHMAN, 1956). The latter value, as the lowest of those enumerated, can be taken as more or less true. On the basis of it and from the value of the specific partial volume of the 80 S ribosomes $\bar{v} = 0.66 - 0.67$ cm^3/g, we obtain a value $w = 1.3$ g/g, i.e., no more than 60% water by weight in the ribosome. An analogous value is also obtained for the 70 S ribosomes of *E. coli*, if we use the least doubtful values of the intrinsic viscosity $[\eta] =$ 5.4—6.1 cm^3/g, obtained by TISSIÈRES et al. (TISSIÈRES et al., 1959); this value gives $w = 1.5 - 1.8$ g/g, i.e., once again about 60% water by weight in the ribosomes. If we make a correction in the calculated values for the deviation of the shape of the ribosomes from the spherical and for the nonideality of its surface, which should also make a contribution to the value of the intrinsic viscosity, the real value for the amount of water retainable in the ribosome proves to be even lower.

HART's data (HART, 1962) on the electron microscopy of frozen-dried ribosomal particles of *E. coli* also lead to a comparatively small value of the swelling; according to the author's estimate, $w = 0.9$ g/g, which means less than 50% water by weight. The data of LANGRIDGE and HOLMES (LANGRIDGE, HOLMES, 1962) on the X-ray diffraction analysis of moist gels of ribosomal particles of *E. coli* also more likely agree with a value of about 50% swelling.

On the basis of the data and considerations cited above, we might think that evidently the actual amount of immobilized water should comprise 50% or even less of the weight of the ribosomal particle, considered in aqueous medium. This corresponds to no more than 60% swelling by volume. It is evident (PRIVALOV, MREVLISHVILI, 1967; PRIVALOV, 1968) that the thermodynamically bound water of hydration may account for no more than 0.5 g per g of the dry weight of the ribosomes so that it may be that about half the water in the ribosome, i.e., 0.5 g per g of dry weight, should comprise of liquid-drop water, mechanically retained in the pores, cavities, and surface folds of the ribosome. We should say that this is a rather low value for the amount of water retainable in the ribosome, contradicting the widespread view of the ribosome as an extremely porous and loose — "spongy" — structure. (It might be mentioned that protein crystals, where there is a close packing of the hydrated protein molecules, and the free water occupies only minimal spaces between them, also contain about 50% water as a whole.) Most likely the ribosome should be considered as a rather compact, closely packed structure, with minimal spaces for water, and evidently one cannot speak of any great looseness of the ribosomes. A study of other properties of the ribosomes in solution also compels us to accept the model of the ribosome as a nonpermeable and in this sense a *rather compact* particle (INOUYE et al., 1963), *with a large amount of intraribosomal interactions* [LERMAN et al., 1966; GAVRILOVA et al., 1966 (1, 2); see below — Section IV, 2, 3, 4, 5).

3. Subdivision into Two Unequal Subparticles ("Subunits")

An attentive study of the morphology of the ribosomes from *E. coli* by electron microscopy indicates that the 70 S particle, as has already been noted, is somewhat prolate in one direction. In the case of negative staining of these particles, it is found that approximately perpendicular to the long axis of the particle there is a groove or cleft, filled with the stain, which subdivides the ribosome into two unequal portions (HUXLEY, ZUBAY, 1960) (see Fig. 11). In a study of the 80 S ribosomes of animals and higher plants, a similar cleft can be detected, although with greater difficulty (SHELTON, KUFF, 1966; ODINTSOVA et al., 1967) (see Fig. 12). It is found that this cleft, detected on the surface of the ribosome, reflects a real subdivision of the ribosomal particle into *two unequal* component *subparticles* (or, as is customarily stated,

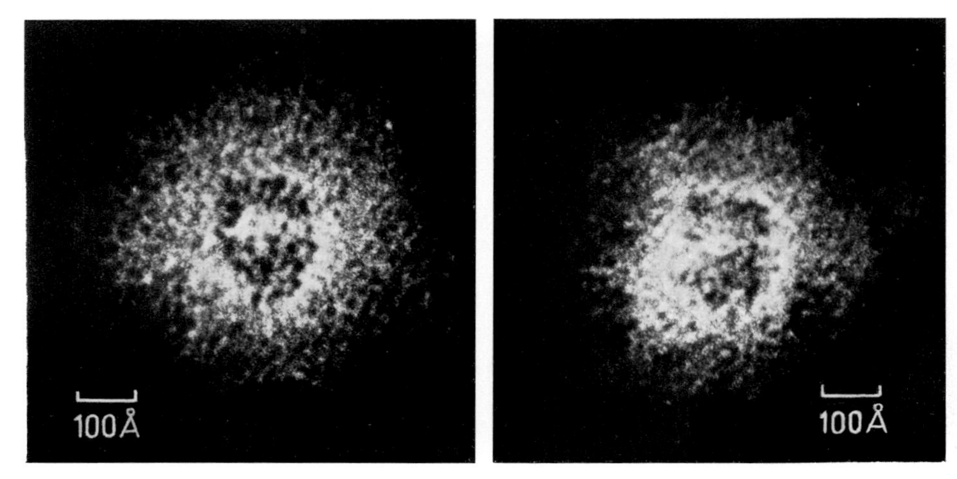

Fig. 11. Electron micrographs of single 70 S ribosomes of *Escherichia coli*, negatively stained with phosphotungstic acid (performed by V. I. BRUSKOV, Institute of Crystallography, Academy of Sciences of the USSR, Moscow). Magnification ×800,000

subunits). A change in the ionic composition of the medium in a suspension of ribosomes, especially a decrease in the concentration of Mg^{++} ions or a substantial increase in the concentration of monovalent cations, leads to dissociation of the ribosome into these two subparticles (see below, Section IV, 1). The dissociation is reversible, and the two unequal subparticles can recombine (reassociate). The construction of the ribosome from two unequal subparticles was first detected precisely in experiments on the dissociation and reassociation of the ribosomes in solution depending upon the ionic surroundings — first for the 80 S ribosomes of yeast (CHAO, 1957), and then very reliably and definitively for the 70 S ribosomes of *E. coli* (TISSIÈRES, WATSON, 1958; TISSIÈRES et al., 1959). The applicability of this scheme of construction of the ribosomes from two unequal subparticles to all ribosomes, independent of the biological source, has sometimes been doubted by certain researchers, but subsequently a more accurate study has invariably confirmed it. At present there is almost no doubt that the construction of the ribosome from two unequal subparticles is a universal principle.

The subparticles into which the 70 S ribosome of *E. coli* dissociates possess sedimentation coefficients $S^0_{20,w} = 50$ S and 31 S (TISSIÈRES et al., 1959); they are denoted as 50 S and 30 S particles, respectively. Their molecular weight, calculated from sedimentation and diffusion and from sedimentation and viscosity, is equal to 1.8×10^6 and $0.7 - 1.0 \times 10^6$ for the 50 S and 30 S particles, respectively (TISSIÈRES et al., 1959).

It is evident that the ratio of their weights is equal to two, i.e., the 50 S particle is twice as heavy as the 30 S particle. In such a case, *one third of the original 70 S ribosome is accounted for by the smaller subparticle (30 S) and two thirds by the larger subparticle (50 S).*

The shape of the 50 S and 30 S subparticles of *E. coli*, when they exist in the dissociated state, has been studied in detail by electron microscopy (HALL, SLAYTER, 1959; HUXLEY, ZUBAY, 1960; HART, 1962). The larger (50 S) subparticle in the dried state possesses approximately spherical outlines; its diameter is about 140 to 170 Å.

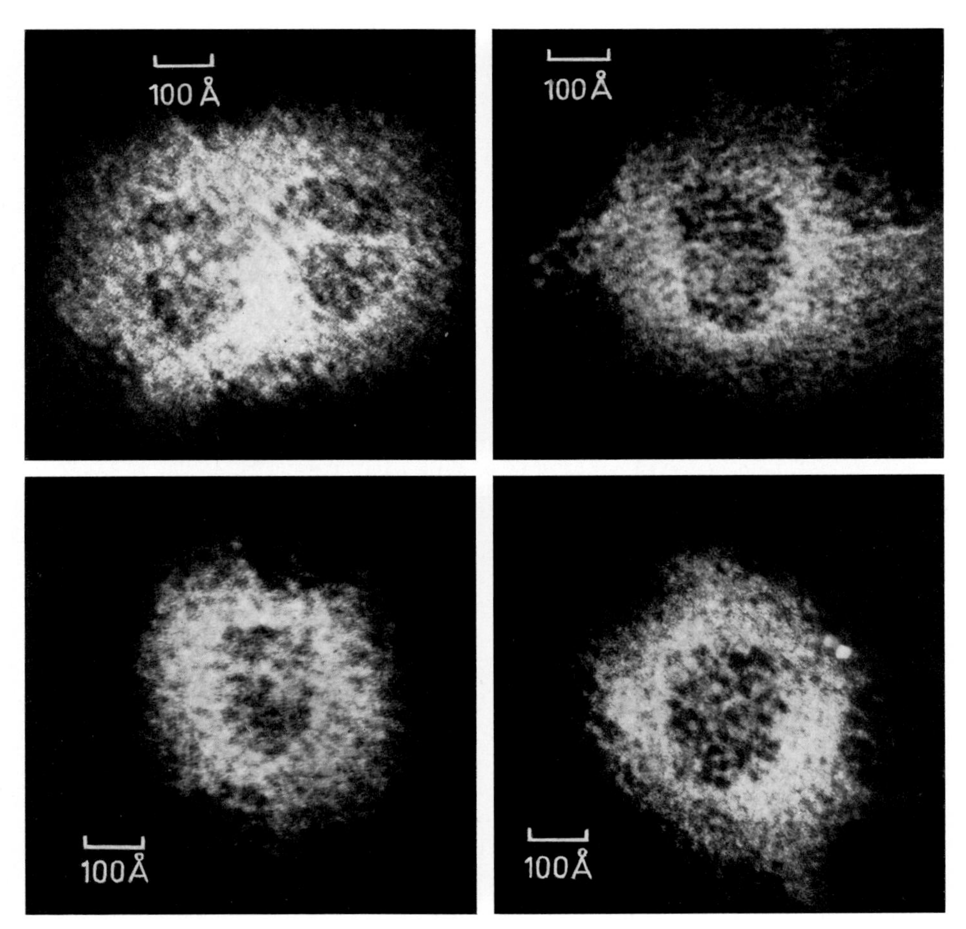

Fig. 12. Electron micrographs of single 80 S bean ribosomes, negatively stained with phosphotungstic acid (performed by V. I. BRUSKOV, Institute of Crystallography, Academy of Sciences of the USSR, Moscow). Magnification ×800,000

It can be more accurately approximated by a slightly oblate ellipsoid of revolution with dimensions of $160 \times 160 \times 140$ Å or $170 \times 170 \times 140$ Å, which gives a "dry" volume of about 2×10^6 Å^3. Under the assumption of 50% (by weight) swelling of the particle ($w \sim 1$ g/g), its effective volume in solution should increase to 5×10^6 Å^3, while its dimensions evidently should increase to $210 \times 210 \times 200$ Å.

Actually, in an electron microscopic study of frozen-dried 50 S particles, their dimensions proved to be substantially greater than those of air-dried 50 S particles: they comprised $230 \times 230 \times 160$ Å (volume 4.6×10^6 Å^3) (HART, 1962). The large volume in this case was due to the fact that in the case of the usual drying of the ribosomes in air, losing water, they become compressed, while in the case of drying from the frozen state, they may retain their original dimensions. Hence, on the basis of the difference of the theoretical "dry" volume, or the volume of air-dried 50 S particles, on the one hand, and the apparent volume of the frozen-dried 50 S particles, on the other hand, the internal swelling of the ribosome was estimated at $w = 0.9$ g/g. This value is in

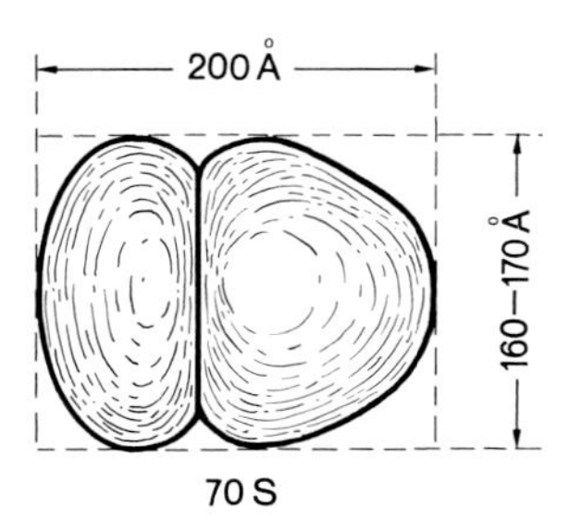

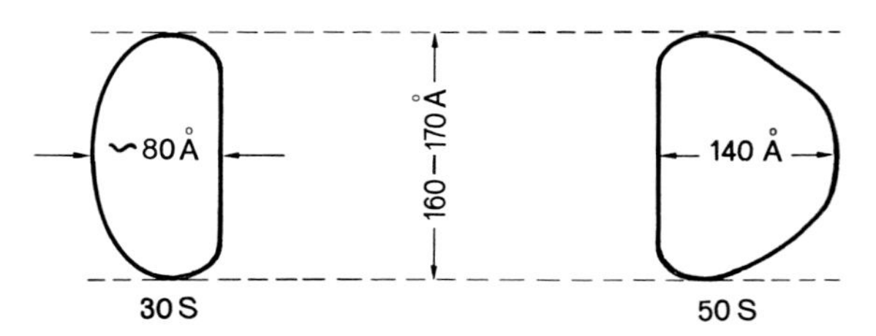

Fig. 13. Scheme of construction of the bacterial 70 S ribosome from two unequal subunits, 30 S and 50 S, and size of the particles in the dried state

rather good agreement with the estimate of the total swelling of the ribosome, made on the basis of measurements of the intrinsic viscosity (see above, Section I, 2).

In contrast to the more or less isometric 50 S particle, the smaller (30 S) ribosomal subparticle in the free form looks flattened in one direction on the electron micrograph, with a deformable structure, frequently with irregular outlines (HALL, SLAYTER, 1959; HUXLEY, ZUBAY, 1960). The height of such a particle, lying with its flattened side on the supporting film, is 70 to 95 Å, while the visible diameter is an average of 150 to 180 Å. If we approximate this particle by an oblate ellipsoid of revolution or by a disc with dimensions of $170 \times 170 \times 80$ Å (ratio of semiaxes about 1:2), then we obtain a dry volume of about 10^6 Å^3. This is approximately just one third of the "dry" volume of the 70 S ribosome and half the "dry" volume of the 50 S subparticle. If the swelling of the ribosomal particles in solution is about $w = 1$ g/g (50% by weight), then the effective volume of the 30 S subparticle in solution should be correspondingly greater, comprising about 2.5×10^6 Å^3; the dimensions of the swollen particle may be assumed equal to $110 \times 210 \times 210$ Å.

In the 70 S ribosome, the two subparticles are combined in such a way that their short axes lie on a single straight line, their sum representing the long axis of the 70 S particle. In this case, the subparticles fit closely to one another, ensuring mutual contact over a large surface. The side of the 50 S particle that is turned to the 30 S particle is flattened, so that the 50 S particle resembles a dome, placed with its base on the 30 S particle. In turn, the 30 S particle is also flattened or even somewhat concave on the side adjoining the 50 S particle, thus effectively fitting the "base" of the 50 S particle. The shape of the subparticles forming the 70 S ribosome and their dimensions are presented schematically in Fig. 13.

Evidently in the complete (70 S) ribosome, the shape of its subparticles is far more fixed and specific than in the dissociated form. This is especially evident in an examination of the 30 S particle: polymorphic and flattened, when taken separately, it acquires a specific, slightly concave-convex shape during association with the 50 S subparticle.

As to the details of morphology of the 70 S ribosomes and their subparticles revealed by electron microscopy and characterizing fine surface "relief" of the particles, one can note in the first place the following: 1. there are parallel stripes on the surface of ribosomal particles, with a periodicity (distances between parallel stripes) of about 30 to 40 Å (SPIRIN, KISSELEV, 1964; HART, 1965); 2. a hollow with a diameter of about 40 Å can be observed on the boundary between the subparticles (BRUSKOV, KISSELEV, 1968).

Although it is also known that the 80 S ribosomes of higher organisms are subdivided into two unequal subparticles, and under definite conditions they dissociate reversibly into these subparticles, there is no such detailed information available as with respect to the 70 S ribosomes. It can be stated definitely only that in the 80 S ribosomes the cleft between the subparticles is far less pronounced, and that the subparticles are considerably more strongly associated to one another than in the case of 70 S ribosomes. However, a *dissociation of the 80 S ribosome into the larger and the smaller subparticles with sedimentation coefficients about 60 S and about 40 S, respectively* can nevertheless be obtained by certain methods (see Section IV, 1). Henceforth we shall denote them as the 60 S and 40 S subparticles of the 80 S ribosome.

TASHIRO and YPHANTIS (TASHIRO, YPHANTIS, 1965), using the sedimentation equilibrium method to study ribosomal particles from guinea pig liver, and PETERMANN and PAVLOVEC (PETERMANN, PAVLOVEC, 1966), on the basis of data on the sedimentation and diffusion of the particles from rat liver sarcoma, determined the molecular weight of the larger subparticle and found that it is equal to $3.5 \times 10^6 - 2.7 \times 10^6$ (about 3×10^6 on the average). Since the molecular weight of the entire 80 S ribosome, according to their data, is $5.2 \times 10^6 - 4.3 \times 10^6$, a molecular weight of $1.8 \times 10^6 - 1.6 \times 10^6$ (an average of 1.7×10^6) remains for the smaller subparticle.

4. Chemical Composition: RNA and Protein Content

a) 70 S Ribosomes

Chemical analysis of preparations of the 70 S ribosomes and their 50 S and 30 S subparticles from *Escherichia coli* shows that they consist solely of RNA and protein, within the limits of the error of determination ± 3 to 4% (TISSIÈRES et al., 1959). The amount of RNA in the ribosomes of *E. coli* substantially exceeds the amount of protein. An analysis of purified preparations gives a value of 60 to 64% for RNA and 40 to 36% for protein, respectively. The usually assumed values for the ribosomes of *E. coli* are 63% RNA and 37% protein. Thus, the RNA/protein weight ratio is about 1.6 – 1.8. The RNA/protein ratio is the same for both subparticles of the ribosome, within the limits of measurement error.

The experimentally determined partial specific volume ($\bar{v}$) of the ribosomes of *E. coli* agrees with the indicated data. Both for the 70 S particles and separately for the 50 S and 30 S subparticles it was close or identical and equal to $\bar{v} = 0.64$ cm^3/g (TISSIÈRES et al., 1959). A theoretical calculation of the partial specific volume of the ribosomes of *E. coli*, performed on the basis of the composition of the ribosomes (63% RNA and 37% protein) and the known partial specific volume of the ribosomal RNA ($\bar{v} = 0.57 \pm 0.02$ cm^3/g; KURLAND, 1960) and protein ($\bar{v} = 0.74$ cm^3/g; SPAHR, 1962) gives a similar value: $\bar{v} = 0.63$ cm^3/g (SPAHR, 1962).

An analysis of ribosome preparations from other bacteria indicates, most likely, a universality of the RNA/protein ratio in all the ribosomes of the 70 S type, independent of the bacterial species. For the pure 50 S subparticles of ribosomes of a species far from *E. coli* — *Streptococcus pyogenes* — the RNA content was found to be 62 to 64% (HESS, HORN, 1964). Even if bacterial species differing sharply from *E. coli* in "ecology" — for example, the extremely halophilic bacterium *Halobacterium cutirubrum* — are taken, here also a chemical analysis of the preparations shows 60% RNA and 40% protein (BAYLEY and KUSHNER, 1966).

b) 80 S Ribosomes

Preparations of the 80 S-type ribosomes, isolated from fungi, higher plants, or animals, usually exhibit substantially lower RNA/protein ratios. In the very first investigations conducted on preparations of yeast ribosomes, values of 40 to 44% were obtained for the RNA content; the remaining 60 to 56% was accounted for by protein; no lipids were detected; the partial specific volume was 0.67 cm^3/g (CHAO, SCHACHMAN, 1956). A value of about 40% RNA was also obtained for ribosome preparations from pea seedlings; a figure of 0.66 cm^3/g was found for the partial specific volume (TS'O et al., 1956, 1958). Originally a value of the RNA content of about 40% was also reported for preparations of 80 S ribosomes from rat liver; protein in these preparations comprised 50 to 55%, while the remaining 5 to 10% was accounted for by lipid (PETERMANN, HAMILTON, 1957; HAMILTON, PETERMANN, 1959). Subsequently, however, as a result of better purification, liver ribosome preparations were evidently freed of admixtures of lipid-containing components; the RNA content in such preparations comprised 45%, protein — 55%; the specific partial volume was experimentally determined as $\bar{v} = 0.664$ cm^3/g (HAMILTON et al., 1962). Evidently, as a general rule, in the case of thorough purification of preparations of 80 S ribosomes from protein and lipid-containing impurities, the RNA/protein ratio approaches 1. Such an RNA/protein ratio — 1.0 ± 0.05 — was obtained in careful analyses of ribosome preparations from rabbit reticulocytes (TS'O, VINOGRAD, 1961). Ribosome preparations containing about 50% RNA were also obtained from rat liver tumors — Jensen sarcoma (PETERMANN, 1960; PETERMANN, PAVLOVEC, 1966) and from Novikoff hepatoma (KUFF, ZEIGEL, 1960); in both cases sodium deoxycholate was used to purify the preparations from membrane components. It is interesting that the parallel isolation of free ribosomes from Novikoff hepatoma without the use of deoxycholate gave preparations with a lower RNA content — 37% (KUFF, ZEIGEL, 1960). In certain cases the isolation of ribosome preparations with a higher RNA content from animal tissues has been reported. For example, in preparations obtained from guinea pig liver, using deoxycholate, determinations gave 56.4 ±2.1% RNA (TASHIRO, SIEKEVITZ, 1965). Deoxycholate-treated ribosomes from calf thymus showed even higher values of the RNA content — 58 to 62%; at the same time, despite two treatments with deoxycholate, they contained 9 to 11% lipids (HESS, LAGG, 1963).

Notwithstanding a certain contradictoriness of data on the chemical composition of the 80 S ribosomes, we can still attempt to formulate some general conclusions on the basis of an analysis of all the available information.

1. The RNA/protein ratio in the 80 S ribosomes is evidently lower than in the bacterial 70 S ribosomes, and is about 1 for the purest preparations. The specific partial volume of these ribosomes is equal to $\bar{v} = 0.66 - 0.67$ cm³/g.

2. Purified 80 S ribosomes, just like bacterial 70 S ribosomes, represent pure ribonucleoprotein, i.e., consist of approximately equal amounts of RNA and protein and do not contain lipids and polysaccharides as intrinsic components. However, in a number of cases preparations of 80 S ribosomes may contain a lipid-containing component bound to the particles; this component can be removed by one means or another without disturbing the structural integrity of the particles, and without any loss of their biological activity.

3. In many cases, if not always, treatment of the 80 S ribosomes with deoxycholate or other detergents, while preserving the essential structural integrity of the particles, may lead to removal of a small portion of the loosely bound ribosomal protein. The RNA/protein ratio in such ribosomes, treated with deoxycholate, increases to 1.2 to 1.5, while the sedimentation coefficient $S^0_{20,w}$ is slightly lowered, to 74 – 79 S.

In those cases when the RNA/protein ratio was determined separately for the two subparticles of the 80 S ribosomes after dissociation of the latter — for example, for the 60 S and 40 S subparticles of the ribosomes of rabbit reticulocytes (TS'O, VINOGRAD, 1961) — it proved to be the same, and just as in whole ribosomes, it was about 1, i.e., the 60 S and 40 S subparticles did not differ appreciably in their RNA and protein content.

There are, however, indications that the smaller (40 S) subparticle of an animal 80 S ribosome has an essentially less buoyant density in CsCl, then the larger (60 S) subparticle (PERRY and KELLY, 1966; INFANTE and NEMER, 1968; BELITSINA et al., 1968); it indicates most probably a somewhat greater relative protein content in the 40 S subparticle in comparison with the 60 S subparticle.

5. Chemical Composition: Bound Low-Molecular Weight Cations

The most important low-molecular weight component of any ribosome are the divalent metal ions, chiefly magnesium ions. The absolute need for ions of divalent metals, primarily magnesium ions, to preserve the structural integrity of the ribosome was recognized in the very first attempts of isolating and physicochemical study of ribosome preparations (CHAO, SCHACHMAN, 1956; CHAO, 1957; TISSIÈRES,

WATSON, 1958; Ts'o et al., 1958; HAMILTON, PETERMANN, 1959; see also the collection: "Microsomal Particles and Protein Synthesis," Pergamon Press, 1958). The gradual removal of Mg^{++} from the ribosomes at first led to the dissociation into subparticles, and then to an irreversible breakdown of their structure. This was most often accompanied by enzymatic degradation of the ribosomal RNA, since as a result of a breakdown of the structure of the subparticles it became very sensitive to the action of nuclease impurities, which are almost alway present in ribosome preparations. As a result, from the very first studies it became clear that Mg^{++} in the ribosome is essential not only for the association of the subparticles with one another, but also to maintain the structure of the subparticles themselves.

In addition to Mg^{++}, certain amounts of Ca^{++} are sometimes detected in ribosome preparations (for example, Ts'o et al., 1958). Ca^{++} ions, added to the ribosome suspension, as has been shown, also stabilize the ribosomes and may replace Mg^{++} to some degree (CHAO, 1957; ELSON, 1959, 1961; HAMILTON, PETERMANN, 1959; CHOI, CARR, 1967). Other alkaline earth cations — Sr^{++}, Ba^{++}, Be^{++} — are either entirely ineffective in maintaining the association of the subparticles in the ribosome when they are added to the medium or effective only to a very slight degree (Sr^{++}). There are indications of a certain stabilizing role of Mn^{++} and Co^{++} (LYTTLETON, 1960, 1962; ABDUL-NOUR, WEBSTER, 1960).

In conclusion analyses of ribosome preparations show that the main metallic cation in them is invariably Mg^{++}; Ca^{++} may be present in certain amounts, while although other cations may sometimes be components of naturally isolated ribosomal particles, they are evidently present only in very small amounts, as "trace elements".

The first determinations of the content of bound metallic cations were performed on preparations of ribosomes from pea seedlings (Ts'o et al., 1958). For preparations of nondissociated 80 S ribosomes, washed with water, they gave values of about 0.30 – 0.33 micromole of divalent metal ions per micromole of phosphorus in the ribosome; this included 0.25 to 0.29 micromole of Mg^{++} and only about 0.04 micromole of Ca^{++}. This means that such ribosome preparations contain about 1600 to 1800 atoms of Mg^{++} per 80 S ribosome, or about 10 mg of Mg^{++} per g of dry weight of the ribosomes (about 1% of the dry weight of the ribosomal particle).

Similar values — 0.2 to 0.3 micromole of Mg^{++} per micromole of phosphorus in nondissociated 80 S ribosomes — were found for washed ribosomal preparations from yeast (LANSINK, 1964) and from rabbit reticulocytes (EDELMAN et al., 1960). Lower values — 0.1 to 0.15 micromole of Mg^{++} per micromole of phosphorus — were obtained for the 80 S ribosomes of frog embryos (BROWN, CASTON, 1962), guinea pig pancreas (SIEKEVITZ, PALADE, 1960) and rat hepatic tumor (Jensen sarcoma) (PETERMANN, 1960).

On the contrary, for nondissociated 70 S ribosomes of *E. coli* (within the Mg^{++} concentration range from 0.005 M to 0.015 M), the Mg^{++} content is appreciably higher than for yeast ribosomes, being of about 0.5 micromole of Mg^{++} per micromole of phosphorus (RODGERS, 1964; GOLDBERG, 1966; CHOI, CARR, 1967); this corresponds to about 2500 atoms of Mg per 70 S ribosome or about 20 mg Mg^{++} per g of dry weight of the ribosomes of *E. coli* (*ca.* 2%).

Upon incubation in media containing a rather large concentration of Mg^{++} ions (0.01 M) animal and plant 80 S ribosomes also can readily bind up to 0.5 micromole of Mg^{++} per micromole of phosphorus (PETERMANN, 1960; LANSINK, 1964). This means that practically all the ribosomal RNA may be present in the ribosome in the form of Mg^{++} salt.

When the Mg^{++} content is lowered below a certain value the ribosomes may dissociate into subparticles. The critical content of Mg^{++} in the ribosomes, below which they dissociate into subparticles, may differ for ribosomes isolated from different biological sources and taken in different functional states. This critical Mg^{++} content is an index of the stability of association of the ribosomal subparticles in the whole particle: the lower this critical concentration, evidently the more firmly the subparticles are bound to one another. In this sense, the two groups of ribosomes, 70 S and 80 S, differ distinctly in the stability of the association of their subparticles: the former begin to dissociate even upon lowering Mg^{++} content below 0.5 micromole per micromole of phosphorus, while for the dissociation of the 80 S particles, the Mg^{++} content must usually be lowered to 0.3 – 0.1 micromole per micromole of phosphorus. This index, the critical Mg^{++} content, below which given ribosomes dissociate into subparticles — may be denoted as the *first critical level of* Mg^{++} *content in ribosomal particles.*

The stability of the ribosomal subparticles themselves is maintained at a substantially lower Mg^{++} content. Thus, if yeast 80 S ribosomes are dialyzed against buffer without Mg^{++}, then simultaneously with dissociation into subparticles they lose their endogenous Mg^{++} to a final content of about 0.15 micromole per micromole of phosphorus (LANSINK, 1964). This residual Mg^{++} is extremely firmly retained by the subparticles themselves. If this content is reduced even more by treatment with chelating agents (EDTA), then the native structure of the subparticles begins to break down.

The same occurs in the 70 S ribosomes of *E. coli*, but at different values of the Mg^{++} content. The dissociation of the ribosomes into 50 S and 30 S subparticles is accompanied by a drop in the Mg^{++} content from 0.5 micromole per micromole of phosphorus to 0.35 – 0.33 micromole per micromole of phosphorus (RODGERS, 1964; CHOI, CARR, 1967). The residual Mg^{++} is far more firmly retained by the subparticles. If, however, one begins to remove it, then further structural transformations (unfolding) occur in the subparticles (RODGERS, 1964; GESTELAND, 1966; see below, Section IV, 2).

The minimum Mg^{++} concentration in the particles that is still sufficient to maintain the original integrity (compactness) of the subparticles may be denoted as the *second critical level of* Mg^{++} *content in ribosomal particles.* The second critical level of Mg^{++} content in ribosomal particles, below which they lose their original compactness, as a rule, is about 0.15 micromole of Mg^{++} per micromole of phosphorus for 80 S ribosomes and about 0.3 micromole of Mg^{++} per micromole of phosphorus for 70 S ribosomes, or something in the region of 1000 Mg atoms per whole ribosome, i.e., about 5 to 10 mg of Mg^{++} per g of dry weight of the ribosomes (0.5 to 1%). However, this level is not constant and also depends greatly, in particular, upon the content of monovalent cations in the medium and on the presence of polyamines within the ribosome. For example, the presence of K^+ ions in the medium somewhat stabilizes the ribosomal particle, and the second critical level of Mg^{++} content may be appreciably lower in the presence of K^+. The removal of polyamines (see below) from the ribosome increases the necessary critical level of Mg^{++} content. Substantial variations may be observed depending upon the biological source of the ribosomes.

Important information on the nature and strength of the binding of Mg^{++} ions in the ribosomal particles is given by experiments on the rate of exchange of bound Mg^{++} with the exogenous Mg^{++} in the surrounding solution. It is found that, as was shown on ribosomes of *E. coli* using radioactive Mg^{++} (Rodgers, 1964), about 35% of the Mg^{++} in the 70 S ribosome exchanges fairly quickly with free Mg^{++} of the medium (in less than 2 min at 0°). This is precisely the labilely bound Mg^{++}, which determines the first critical level necessary for the maintenance of association of the subparticles in the 70 S particle. The remaining Mg^{++} of the ribosome, maintaining the compactness and integrity of the subparticles themselves, practically does not exchange with the free Mg^{++} in a short period of time; the rate of its exchange does not exceed 5 to 6% per hour at 0°. The strength of its binding in the subparticles is indicated by the fact that simple, even prolonged dialysis against magnesium-free buffer usually does not suffice for its removal; instead, the use of either chelating agents of the type of EDTA (Gesteland, 1966) or of high (0.5 M and above) competitive concentrations of monovalent cations in the medium [Spirin et al., 1963; Gavrilova et al., 1966 (1)] is required.

Thus, it may be assumed that the endogenous Mg^{++} of the ribosomal subparticles, ensuring their integrity and corresponding to the second critical level of Mg^{++} content in the ribosomes, is *firmly bound and slowly exchangeable* with the free Mg^{++} of the medium. The supplementary amount of Mg^{++} that is necessary to maintain association of the subparticles in the ribosome, i.e., the addition of which results in the reaching of the first critical level of Mg^{++} in the ribosome, is represented by *labilely bound and rapidly exchangeable* Mg^{++} ions.

We should emphasize once again, however, that the concept of the critical levels of Mg^{++} content in the ribosomal particles as formulated cannot be considered as absolute. In particular, an appreciable role in maintaining the integrity of the subparticles and in their association into the complete particle, may also be played by organic cations, both di- and polyamines, partially replacing the role of Mg^{++}. Therefore, depending upon their content in the ribosomes, the critical level of the Mg^{++} content may evidently differ and vary somethat even for ribosomes from the same species.

Polyamines, as a component of the ribosomes, have been studied most detailly in preparations of ribosomes from *E. coli*. Both the qualitative assortment and their quantitative content vary somewhat according to the data of different authors. It is quite probable that this may reflect real variations depending upon the conditions of culturing, phase of growth, strain, etc., i.e., that this component of the ribosome is not so constant, not so strictly fixed in quantity and composition as others. Thus, Zillig et al. (Zillig et al., 1959) detected diamines — *putrescine, cadaverine, and 1,3-diaminopropane* — in ribosomes of *E. coli*; the first two predominated, and were present in a 2:1 ratio; spermidine was not detected; the total diamine content was 2.5% of the dry weight of the ribosomes. Cohen and Lichtenstein (Cohen, Lichtenstein, 1960) found 1.5% polyamines in the ribosomes of *E. coli*, and only putrescine and *spermidine* in a 3:1 to 2:1 ratio. Spahr (Spahr, 1962) determined the polyamine content in the ribosomes of *E. coli* as equal to only 0.4% of the dry weight of the ribosomes, detecting putrescine, spermidine, and cadaverine, in a 3:2:1 ratio, along with traces of *spermine*. In preparations of animal ribosomes, as a rule, spermine is detected, and cadaverine and 1,3-diaminopropane have also been found (Zillig et al., 1959;

SIEKEVITZ, PALADE, 1962). Spermidine, putrescine, and a little spermine have also been found in yeast ribosomes (OHTAKA, UCHIDA, 1963).

All the enumerated poly- and diamines, added in small concentrations to the solution, have been found to appreciably stabilize the ribosomes against enzymatic and thermal degradation and prevent dissociation into subparticles in buffers with a low Mg^{++} content and even without any Mg^{++} at all; spermidine is especially effective in this respect. There are convincing data indicating that the addition of polyamines permits an appreciable lowering of the optimum concentration of Mg^{++} in a ribosome suspension during work with cell-free protein-synthesizing systems, and also leads to an increase in the activity of such systems (MARTIN, AMES, 1962; NATHANS, LIPMANN, 1961; LUCAS et al., 1963; BRETTHAUER et al., 1963; HERSHKO et al., 1961). Consequently, polyamines may represent an important component of the ribosome, evidently not simply supplementing the stabilizing role of Mg^{++}, but also playing their own intrinsic regulating role in the creation of the ribosomal structure best adapted for the performance of its functions.

6. Summary

The ribosomes are compact particles which may be approximated by a slightly prolate ellipsoid of revolution with an axial ratio not exceeding 1.5. The ribosomes of bacteria and blue-green algae possess a molecular weight of about 2.8 to 3×10^6 and a sedimentation coefficient of about 70 S; the dimensions of the dry 70 S ribosomes are $200 \times 170 \times 170$ Å. The ribosomes of animals, higher plants, fungi, and algae are somewhat larger; they are characterized by a molecular weight of 4 to 5×10^6, a sedimentation coefficient of about 80 S, and dimensions in the dry state of $240 \times 200 \times 200$ Å. In aqueous medium, the linear dimensions of the ribosomes are evidently 20 to 40% greater than in the dry state. The total amount of water retainable by the ribosome evidently is relatively small, comprising about 1 g of water per g of dry weight, or even less, i.e., the swelling of the ribosomes does not exceed 60% by volume.

A universal structural feature of all ribosomes is that they are constructed from two unequal subparticles. The 70 S ribosome is subdivided into 50 S and 30 S subparticles, with a 2:1 weight ratio. The 80 S ribosome is analogously subdivided into 60 S and 40 S subparticles.

In chemical composition, the ribosome represents a ribonucleoprotein and consists almost entirely of ribosomal RNA and structural ribosomal protein. The RNA: protein ratio in bacterial 70 S ribosomes is 1.6 to 1.8. In the 80 S ribosomes of higher organisms, there is relatively more protein; the RNA: protein ratio in them is close to 1.

Cations of divalent metals, chiefly Mg^{++} and in a number of cases perhaps Ca^{++} as well, represent a necessary low-molecular weight component of all ribosomes. Their content may reach 0.5 micromole per micromole of phosphorus (2500 to 3000 Mg atoms per ribosome, or 2 to 2.5% of the dry weight) so that a predominant part or almost all of the RNA in the ribosome may exist in the form of Mg^{++} salt. When the Mg^{++} content in the ribosome is reduced, it at first dissociates into subparticles when a definite level is reached, and then, if the Mg^{++} content drops below 1000 atoms per complete ribosome (1 to 0.5% of the dry weight), the original compactness and integrity of the particles is disrupted. Another important cationic component, which evidently also takes part in the stabilization of the ribosomal structure, is polyamines — spermidine, spermine, putrescine, and cadaverine; their content in bacterial ribosomes is up to 0.4 to 2.5% of the dry weight.

II. Ribosomal RNA

1. Molecular Weight Characteristics

As has already been mentioned, ribosomal RNA comprises no less than half of the dry weight of the ribosomal particles and is thus their most important structural component. According to the sedimentation and molecular weight characteristics, the ribosomal RNA of any cell is subdivided into *three discrete classes* of molecules: 1. high-molecular weight 23 S – 28 S RNA (molecular weight more than 1×10^6); 2. high-molecular weight 16 S – 18 S RNA (molecular weight less than 1×10^6); and 3. relatively low-molecular weight 5 S RNA (molecular weight about 40,000).

High-Molecular Weight Ribosomal RNA. The presence of two discrete classes of molecules of high-molecular weight ribosomal RNA is due to the subdivision of the ribosome into two unequal subparticles. Each subparticle contains *one* molecule of high-molecular weight ribosomal RNA. Consequently, the larger (50 S or 60 S) subparticle contains a molecule of 23 S – 28 S RNA, while the smaller (30 S or 40 S) contains a molecule of 16 S – 18 S RNA.

In the case of *bacterial ribosomes*, the 50 S subparticle contains one molecule of ribosomal RNA with a molecular weight of about 1.1×10^6; the 30 S subparticle contains one RNA molecule with a molecular weight of about 0.55×10^6 (KURLAND, 1960; GREEN, HALL, 1961; STANLEY, BOCK, 1965). Usually, according to KURLAND, their sedimentation coefficients are assumed equal to 23 S and 16 S, respectively; therefore, the high-molecular weight ribosomal RNA's of bacteria are customarily denoted as *23 S and 16 S RNA*. However, it should not be forgotten that this is to some degree arbitrary. Actually, the values of the sedimentation coefficients of RNA depend substantially upon the ionic conditions in solution, and different authors by no means use any standardized solvent. Moreover, the value of the sedimentation coefficients of RNA may vary appreciably depending upon impurities of bound ions of divalent and polyvalent metals in the RNA preparation itself. Evidently one of the best standard solvents, which to some degree eliminates the uncontrollable influence of impurity cations, is 0.1 M NaCl with 0.01 M EDTA. As a result of sedimentation analysis of ribosomal RNA's from a number of bacterial species in an analytical ultracentrifuge, it was found that they are all characterized by practically the same values of the sedimentation coefficient; *in 0.1 M NaCl with 0.01 M EDTA*, in the presence of 20 µg/ml polyvinyl sulfate, pH 4.5, the sedimentation coefficient ($S^0_{20,w}$) of bacterial RNA (including *E. coli*) was about *20 S (20.4 ± 0.6 S)*, and about *15 S (15.2 ± 0.4 S)* for the large and small components, respectively[1]; in 0.01 M MgCl, the sedimentation coefficients increased to 28 S for the large component and

[1] These values were obtained by measuring $S_{20,w}$ at *low RNA concentrations* with ultraviolet optics. Extrapolation of the values of $S_{20,w}$ obtained at high RNA concentrations with schlieren optics gives higher values of $S^0_{20,w}$, up to 23 S and 17 S in an *analogous solvent* (phosphate buffer with ionic strength 0.1, 0.01 M EDTA) (SPIRIN, 1961).

21 S for the small component of high-molecular weight ribosomal RNA (TAYLOR et al., 1967).

The molecular weights, and consequently the sedimentation coefficients of ribosomal RNA from the 80 S ribosomes of animals, fungi and higher plants, i.e., evidently all the Eukaryotes, are somewhat higher. The sedimentation coefficients usually observed by various authors for the *ribosomes of vertebrate animals* are about 28 to 30 S and about 17 to 18 S (GIERER, 1958; CHENG, 1959, 1960; HALL, DOTY, 1959; TS'O, SQUIRES, 1959; LITTAUER, 1961; PETERMANN, PAVLOVEC, 1963, 1966; HENSHAW, 1964; DE BELLIS et al., 1964; BROWN, GURDON, 1964; MONTAGNIER, BELLAMY, 1964; CLICK, TINT, 1967); therefore animal ribosomal RNA's are always arbitrarily denoted as *28 S and 18 S RNA*. Evidently the most probable values of the molecular weights are about 1.7×10^6 for the 28 S RNA and about $0.7 - 0.8 \times 10^6$ for the 18 S RNA of animal ribosomes (see GIERER, 1958; PETERMANN, PAVLOVEC, 1966; HAMILTON, 1967). Unfortunately, there have been no wide comparative investigations of the sedimentation coefficients of animal ribosomal RNA's under standard ionic conditions in the presence of EDTA.

For *invertebrate animals*, in particular insects, the values of the sedimentation coefficients of ribosomal RNA are evidently somewhat lower than for vertebrates (APPLEBAUM et al., 1966).

The cytoplasmic 80 S ribosomes of *higher plants* contain ribosomal RNA that are characterized by somewhat lower sedimentation coefficients, and correspondingly somewhat lower molecular weights than RNA from the animal 80 S ribosomes. Thus, for the ribosomal RNA of the cytoplasm of higher plants, sedimentation coefficients of 25 S and 16 S *(25 S and 16 S RNA)* have been noted (POLLARD, 1964; CLICK, HACKETT, 1966; CLICK, TINT, 1967; STUTZ, NOLL, 1967). Evidently the same sedimentation coefficients are also characteristic of the 80 S ribosomes of lower plants, in particular, fungi and algae (STUTZ, NOLL, 1967). In 0.1 M NaCl and 0.01 M EDTA in the presence of 20 μg/ml polyvinyl sulfate pH 4.5, the exact values of the sedimentation coefficients of ribosomal RNA's of different species of fungi are about 23.4 ± 0.5 S and 16.4 ± 0.2 S, i.e., higher than those of bacterial ribosomal RNA's under the same conditions (TAYLOR et al., 1967). Evidently the cytoplasmic ribosomal RNA's of higher plants under these conditions will exhibit the same sedimentation coefficients, 23 and 16 S. In 0.01 M $MgCl_2$, the sedimentation coefficients of ribosomal RNA's of fungi (and evidently of higher plants as well) increase to 30 S and 20 S (TAYLOR et al., 1967). It may be calculated that the molecular weights of ribosomal RNA of plants should be about $1.3 - 1.5 \times 10^6$ for the 25 S RNA and about $0.6 - 0.7 \times 10^6$ for the 16 S RNA.

It is interesting that the 70 S ribosomes of the *chloroplasts* and *mitochondria* of higher (eukaryotic) organisms contain ribosomal RNA indistinguishable in sedimentation coefficients, and consequently, in all probability in molecular weights as well, from the bacterial ribosomal RNA (23 S and 16 S) (STUTZ, NOLL, 1967; KÜNTZEL, NOLL, 1967).

The data cited above on the sedimentation coefficients and molecular weights of high-molecular weight ribosomal RNA are summarized in Table 2.

Thus, it can be seen that on the whole, in living nature the variations of the molecular weights of high-molecular weight ribosomal RNA are not very great. Within each of the great kingdoms of living nature — in any case, among bacteria,

Table 2. *Sedimentation coefficients and molecular weights of high-molecular weight ribosomal RNA*

Source of RNA	Sedimentation coefficient $S^0_{20,w}$			Mol. weight
	Values usually accepted (ionic strength ~ 0.1)[a]	In 0.1 M NaCl with 0.01 M EDTA, low concentrations of RNA[b]	In 0.01 M $MgCl_2$[b]	
Bacteria (70 S ribosomes)	23 and 16	20 and 15	28 and 21	1.1×10^6 and 0.55×10^6
Chloroplasts of higher plants (70 S ribosomes)	23 and 16	(20 and 15)[c]	—	1.1×10^6 and 0.55×10^6
Cytoplasm of higher plants, fungi, and algae (80 S ribosomes)	25 and 16	23 and 16	33 and 23	$1.3-1.5 \times 10^6$ and 0.6×10^6
Invertebrate animals (80 S ribosomes)	26 and 17	(24 and 17)[c]	—	1.6×10^6 and 0.7×10^6
Vertebrate animals (80 S ribosomes)	28 and 18	(25 and 17)[c]	—	1.7×10^6 and 0.8×10^6

[a] See Kurland, 1960; Stutz, Noll, 1967; Applebaum et al., 1966.
[b] Data of Taylor et al., 1967, for bacteria and fungi.
[c] Values proposed on the basis of a comparison of the data of Taylor et al., 1967, with the values usually accepted.

among fungi, evidently among vertebrate animals, and possibly among higher plants — there are practically no variations in the molecular weights of the ribosomal RNA. This extremely low variability, almost universality, of the ribosomal RNA is also manifested in their other characteristics (see below).

Low-Molecular Weight Ribosomal RNA. In addition to the structural high-molecular weight ribosomal RNA, comparatively low-molecular weight RNA with a sedimentation coefficient of about 5 S has also been detected in the ribosomes both of bacteria and of higher organisms (Rosset, Monier, 1963; Rosset et al., 1964; Comb, Katz, 1964; Comb et al., 1965; Zehavi-Willner, Comb, 1966; Comb, Zehavi-Willner, 1967). Since it is similar to the "soluble" (adaptor) RNA or tRNA (4 S RNA) in the ratio of the four principal bases (G, A, C, U — see below) and in a sedimentation coefficient, it is sometimes called "t-like RNA". However, in contrast to tRNA, the indicated 5 S RNA contains no methylated bases and pseudouridine, possesses no accepting CCA-end and is incapable of enzymatically adding terminal C and A nucleotides, does not participate in the activation and accepting of amino acids and in general exhibits no homology of its nucleotide sequence with that of the tRNA chains. Its molecular weight is appreciably greater than that of tRNA; for example, the 5 S RNA from the ribosomes of *E. coli* possesses a chain of 120 nucleotides (mol. wt. about 40,000) (Brownlee, Sanger, 1967; Brownlee et al., 1967), as against 75 to 85 nucleotides in the 4 S tRNA (mol. wt. about 27,000). The 5 S RNA is not encountered in the free form in the cell, but is always contained in the ribosomes, as a component of the larger (50 S or 60 S) subparticle. When the ribosome dissociates into subparticles, the 5 S RNA remains bound to the larger subparticle, but evidently it is easily removed from it during structural transformations of the unfolding or disassembly type (see below, Section IV, 2, 3) (COMB, SARKAR, 1967; MONIER, 1967). Evidently each ribosome contains one molecule of 5 S RNA. It is very likely that the site of its localization in the larger ribosomal subparticle is situated precisely on the contacting surface, i.e., on the surface joining the smaller subparticle (SARKAR, COMB, 1967).

Thus, it can now be considered that the ribosome — for example, the ribosome of *E. coli* — contains one molecule each of 23 S RNA, 16 S RNA, and 5 S RNA; their relative percentages by weight in the ribosomes of *E. coli* are 65%, 33%, and 2%, respectively (that is, the 5 S RNA accounts for only a little more than 1% of the dry weight of the ribosome).

Nothing is known about the functions of the ribosomal 5 S RNA.

2. Continuity of Polynucleotide Chains

At the end of the fifties to the beginning of the sixties, experimental data were obtained in a number of laboratories, indicating that each molecule of ribosomal RNA consists not of one continuous chain, but of a larger or smaller number of shorter chains (subunits). Thus, a number of authors (HALL, DOTY, 1958, 1959; TAKANAMI, 1960; OSAWA, 1960; TASHIRO et al., 1960; ARONSON, MCCARTHY, 1961; HELMKAMP, TS'O, 1961) have observed the formation of lower molecular weight particles after the heating of solutions of various ribosomal RNA's. This fact was explained by the presence of subunits in the ribosomal RNA macromolecule, bound to one another by "weak" bonds, which are broken upon heating. It was proposed that such "weak" bonds may be hydrogen bonds or divalent and polyvalent metal bridges. Actually, ARONSON and MCCARTHY (ARONSON, MCCARTHY, 1961) observed a breakdown of the ribosomal RNA molecule from *E. coli* to particles with a molecular weight of down to 30,000 as a result of dialysis of a solution of RNA against magnesium-free buffer. On the other hand, HELMKAMP and TS'O (HELMKAMP, TS'O, 1961; TS'O, HELMKAMP, 1961) reported that ribosomal RNA from pea seedlings, characterized originally by two components with sedimentation coefficients 27—28 and 17—18 S, exhibits irreversible dissociation into subunits as a result of cleavage of hydrogen and hydrophobic bonds with formamide or dimethyl sulfoxide, and also as a result of heating of the aqueous solutions; after such treatment and its reconversion to aqueous solution at room temperature, it sediments as a heterogeneous peak with an average sedimentation coefficient of 8—10 S.

Data with opposite results were first obtained in 1960 on animal ribosomal RNA (SPIRIN, MILMAN, 1960; SPIRIN, 1960), and soon after were confirmed on bacterial ribosomal RNA from *E. coli* and plant RNA from pea seedlings (SPIRIN, 1961; BOGDANOVA et al., 1962). In these experiments the authors succeeded in obtaining high-molecular weight ribosomal RNA's that do not dissociate upon heating and when divalent and polyvalent metal cations are removed. All the RNA's studied exhibited a multiple rise in the specific viscosity as a result of thermal unfolding of their polynucleotide chains; the presence of substantial EDTA concentrations did not affect this rise. Cooling of the solutions of ribosomal RNA from *E. coli* after heating, when it is characterized by an unfolded state of the chain (80°), both in the presence and in the absence of EDTA, led to complete restoration of the original dimensions of the molecule, including the large and small components of the ribosomal RNA (23 S and 16 S). Finally, long continuous threads of ribosomal RNA in the unfolded state were detected in electron microscopic studies; the longest threads on micrographs of the ribosomal RNA possessed a length of up to 18,000 Å, which should evidently correspond to one continuous extended polynucleotide chain of the larger component of the ribosomal RNA, with a molecular weight of about 1.1×10^6. Thus, according to these data, each molecule of the ribosomal RNA is evidently consists of a single continuous polynucleotide chain, which is unfolded upon heating, not dissociating and thus giving a substantial thickening (increase of viscosity) of the solution, and which then is readily refolded into the original particle after cooling.

Actually, it is evidently far easier to obtain "dissociation" of the ribosomal RNA molecules than not to obtain it. Such a false picture of "dissociation" is possible in view of the very great vulnerability of the polynucleotide chain of RNA. Special mention should be made of the following possibilities of obtaining artifacts: a) thermal degradation of the polyribonucleotide chain, especially during prolonged heating; b) enzymatic degredation chiefly on account of the presence of traces of the ubiquitous ribonucleases in the RNA preparations; c) the appearance of a few preexisting breaks in the polynucleotide chain, produced during isolation of the ribosomes or RNA preparation from the ribosomes, and not detected at room temperature because of the maintenance of the integrity of the entire compact RNA particle on account of numerous intramolecular "complementary" interactions. In the latter case, when the complementary interactions break down, there actually can be a dissociation of the particle into the individual constituent chains. This has been demonstrated for a number of preparations of viral RNA that have lost their infectivity (Gavrilova et al., 1959; Haselkorn, 1962). The possibility of a disruption of the integrity (continuity) of the polynucleotide chain of RNA within the ribosome, without any visible change in the physicochemical characteristics of these ribosomes, has also been demonstrated (Shakulov et al., 1962). Such a disruption (latent degradation of ribosomal RNA within the ribosomes) can occur during the isolation of the ribosome preparations, and it can be produced artificially, through the action of low concentrations of exogenous ribonuclease upon the ribosomal particles; the sedimentation characteristics of the particles are thereby preserved. Thus, if especially careful efforts are not made to obtain continuous chains, and the opposite viewpoint is used as a basis, then degradation of the polynucleotide chain or the revealing of preexisting breaks of this chain is very easy to demonstrate experimentally and to represent as "dissociation" into supposedly naturally existing subunits.

On the basis of the aforementioned, it might be thought that in the experiments of a number of authors, the so-called dissociation of RNA was actually a consequence either of nuclease contaminations of their preparations or of excessively drastic thermal treatment, leading to degradation of the chain, or poor quality of the preparations themselves, possibly consisting of RNA with a broken chain. Actually, *native* ribosomal RNA molecules represent continuous polynucleotide chains, which, depending upon the molecular weight of the RNA, are made up of 3200—5500 nucleotides for the 23 S—28 S component and of 1600—2500 for the 16 S—18 S component (Spirin, 1960, 1961, 1962; Bogdanova et al., 1962).

As a result of a direct experimental reexamination of the studies supporting the hypothesis of subunits of ribosomal RNA, it was shown that the detection of "subunits" during the heating of solutions of RNA or during the removal of ions of di- and polyvalent metals is an artifact, reflecting the degradation of RNA during the experiment, and the conclusion of continuity of the polynucleotide chains of high-molecular weight ribosomal RNA's was confirmed (MÖLLER, BOEDTKER, 1962; BOEDTKER et al., 1962). This conclusion was also confirmed in a subsequent work, where ribosomal RNA was treated with formamide, dimethylsulfoxide, EDTA, low ionic strength, and heat, and in these cases no dissociation into subunits was detected (STANLEY, BOCK, 1965). Finally, a confirmation of covalent continuity of the polynucleotide chains of high-molecular weight ribosomal RNA was also obtained by a chemical determination of the number of terminal groups [LANE et al., 1963; MIDGLEY, 1965 (1)].

In certain more recent studies, data are reported in favor of the idea that although the covalently-linked chain of high-molecular weight ribosomal RNA is continuous, it contains special "weak points," differing in stability from the usual internucleotide linkages. For example, it is noted that the chain of the 23 S ribosomal RNA of *E. coli* is readily cleaved exactly in the middle under mild alkaline conditions, pH 9 to 9.5, yielding two 16 S fragments [MIDGLEY, 1965 (2)]. However, when we tried to corroborate these experiments in our laboratory, we found that especially carefully deproteinized preparations of RNA evidently do not always give the indicated effect, and the 23 S RNA may be stable at pH 9.5 (0.1 M glycine buffer) for 10 h, at 37° (BARULINA, unpublished).

Furthermore, it is sometimes noted that brief heating of a preparation of RNA or the action of agents of the type of formamide or dimethylsulfoxide can lead to the same breakdown of animal 26 S – 28 S ribosomal RNA in half, into two 16 – 18 S molecules (PETERMANN, PAVLOVEC, 1963, 1966; BROWN, LITTNA, 1964; APPLEBAUM et al., 1966). In the latter study (APPLEBAUM et al., 1966), it was discovered that this process is not a cleavage of a covalent bond, but reflects a real dissociation of non-covalent bonds and indicates a *preexisting* cleavage in the covalent chain of the 26 S RNA. The monodispersity of the dissociation product indicates that the point of cleavage is very specific and always lies in the middle of the chain; in the opinion of the authors, such specificity of the point of preexisting cleavage may be due to the action of nucleases upon RNA while still within the ribosomes, prior to their deproteinization.

In our experiments with ribosomes of *E. coli* and their stripped derivatives (see below, Section IV, 3), we repeatedly observed "spontaneous" cleavage, in half, of the chain of 23 S RNA which was still in the state of ribonucleoprotein particles despite the absence of exogenous ribonuclease activity (BARULINA, BELITSINA, IVANOV, LERMAN, SPIRIN, unpublished). This permits us to believe that such specific degradation of the 23 S RNA chain in half may be due to some catalytic action of an intrinsic structural ribosomal factor (in all probability, a protein), attached to the RNA chain, rather than to exogenous nucleases.

Thus, in all probability, isolated carefully purified ribosomal RNA does not exhibit any special "weak points" in the chain, differing in stability from the usual internucleotide linkages. At the same time, in ribosomal RNA *within ribosomal particles* there are definite preferential points of cleavage, evidently determined by the structure of the particles. Thus, during the "spontaneous" degradation of RNA *within the ribosomes*, the covalent chain of 23 S – 28 S RNA is broken in the first place in half, and as a result, either two 16 S components instead of the one 23 S are isolated from the 50 S subparticle (in the case of *E. coli*), or the 26 – 28 S RNA, isolated from the 60 S subparticle (in the case of animal ribosomes) contains a preexisting cleavage and dissociates readily into two 16 S – 18 S fragments.

3. Nucleotide Composition

a) Proportions of the Four Main Nucleotides

Although the two classes of high-molecular weight ribosomal RNA's are localized in two different subparticles in the ribosome, have a great difference in molecular weight, and also, as is now known, are synthesized in the cell on different DNA regions (different genes) [YANKOFSKY, SPIEGELMAN, 1963; ATTARDI et al., 1965 (1, 2); OISHI, SUEOKA, 1965], they possess very similar proportions of their four main nitrogenous bases (G, A, C, U), i.e., a similar nucleotide composition.

Table 3. *Nucleotide composition of E. coli ribosomal RNA's*

RNA Component	Nucleotide content, mole %				$\frac{G+C}{A+U}$	$\frac{Pu}{Py}$	$\frac{G+U}{A+C}$
	G	A	C	U			
23 S	32.8	25.5	21.0	20.6	1.17	1.40	1.15
16 S	31.6	24.5	22.5	21.3	1.18	1.28	1.12
5 S	33.8	19.2	30.0	17.0	1.76	1.13	1.03

Thus, in *bacterial* ribosomes, either no difference at all can be detected in the nucleotide composition of the 23 S and 16 S RNA's, or the detectable difference is very small and usually consists of a slightly larger content of purines (G, A) in the 23 S RNA (SPAHR, TISSIÈRES, 1959; BOLTON et al., 1959; WOESE, 1961; MIDGLEY, 1962). The low-molecular weight 5 S ribosomal RNA of bacteria, on the contrary, may differ rather substantially in nucleotide composition from the high-molecular weight 23 S and 16 S RNA (ROSSET et al., 1964; SCHLEICH, GOLDSTEIN, 1966). The nucleotide composition of the 23 S, 16 S, and 5 S ribosomal RNA's of *E. coli*, derived as an average from a number of literature data (SPAHR, TISSIÈRES, 1959; BOLTON et al., 1959; MIDGLEY, 1962; ROSSET et al., 1964; BROWNLEE et al., 1967), is presented in Table 3.

The nucleotide composition of ribosomal RNA's of the most different species of bacteria, of the most diversified taxonomic positions, and with the most varied DNA composition, changes very little, exhibiting striking similarity (SPIRIN et al., 1957; BELOZERSKY, SPIRIN, 1958, 1960; WOESE, 1961; MIDGLEY, 1962). All bacterial high-molecular weight ribosomal RNA's are characterized by a small predominance of the sum G + C over the sum A + U [$(G+C):(A+U) \simeq 1.2$]; a substantial predominance of purine nucleotides over the pyrimidines ($Pu:Py \simeq 1.4-1.3$), and an approximate equality of the sums G + U and A + C [$(G+U):(A+C) \simeq 1-1.1$]. (In alkaline hydrolyzates of the RNA preparations, the latter equality frequently deviates from 1 in the direction of larger values — evidently as a result of a certain deamination of C, with its conversion to U; most likely the native composition of ribosomal RNA strictly obeys the indicated equality.)

High-molecular weight RNA's of the cytoplasmic ribosomes of *higher plants* possess a nucleotide composition very similar to the composition of the bacterial ribosomal RNA's. Just as among bacteria, the RNA composition of higher plants evidently varies little from species to species (URYSON, BELOZERSKY, 1959; VANYUSHIN, BELOZERSKY, 1959). Within the class of Dicotyledons, the RNA composition is practically the same (VANYUSHIN, BELOZERSKY, 1959). The nucleotide composition of the larger — 25 S — component of the ribosomal RNA of higher plants (class of Dicotyledons) is practically identical with that of the bacterial RNA's; the composition of the smaller — 16 S — component exhibits a somewhat larger content of U and a smaller content of C (CLICK, HACKETT, 1966; POLLARD, 1964).

In contrast to bacteria and higher plants, among *lower plants*, belonging to the Eukaryotes, there may evidently be more substantial variations of the nucleotide composition of ribosomal RNA's, depending upon the species. In fungi, for example, the composition of the total RNA, evidently reflecting chiefly the composition of the

Table 4. *Nucleotide composition of preparations of mammalian ribosomal RNA's*

RNA Component	Nucleotide content, mole %				$\frac{G+C}{A+U}$	$\frac{Pu}{Py}$	$\frac{G+U}{A+C}$
	G	A	C	U			
28 S	36.1	15.8	31.0	17.2	2.03	1.08	1.14
18 S	30.6	20.3	27.6	21.4	1.40	1.04	1.09
5 S	31.0	17.7	26.4	25.0	1.34	0.95	1.27

ribosomal RNA, may be both of the usual "GC-type" (with a predominance of G and C over A and U) and of the "AU-type" (with a predominance of A and U), but in most species $G+C \simeq A+U$ (URYSON, BELOZERSKY, 1960; VANYUSHIN et al., 1960).

The nucleotide composition of the ribosomal RNA's of *vertebrate animals* differs substantially from the composition of the bacterial ribosomal RNA's, and at the same time, evidently is very similar within the vertebrate group (LERNER et al., 1963; ZIMMERMANN et al., 1963; MONTAGNIER, BELLAMY, 1964; DE BELLIS et al., 1964; BROWN, GURDON, 1964; ELLEM, 1966; HIRSCH, 1966; GAZARYAN, SHUPPE, 1966, 1967; GALIBERT et al., 1966). Here, as a rule, substantial differences are noted in the nucleotide composition of the two classes of high-molecular weight ribosomal RNA: in most cases analysis shows an appreciably larger content of A and U and a smaller content of G and C in preparations of 18 S RNA in comparison with 28 S RNA. The average data on the nucleotide composition of high-molecular weight ribosomal RNA's of mammals (the composition of RNA from different species of mammals evidently is quite indistinguishable) (MONTAGNIER, BELLAMY, 1964; DE BELLIS et al., 1964; ELLEM, 1966; HIRSCH, 1966), as well as data on the composition of mammalian 5 S RNA (GALIBERT et al., 1966), are presented in Table 4.

The composition of the ribosomal RNA's of birds (LERNER et al., 1963) and amphibians (BROWN, GURDON, 1964) differs only slightly, if any, from the composition of mammalian RNA. However, the composition of the RNA of fish may differ significantly from the composition indicated above and may vary strongly from species to species (MEDNIKOV et al., 1965).

It may be probable, however, that the larger content of A and U in preparations of 18 S RNA in comparison with 28 S RNA, ascertained by all authors, does not reflect a true difference, but is a consequence of contamination of the preparation of ribosomal 18 S RNA by a fraction of messenger RNA, possessing the same sedimentation coefficient and characterized by a predominance of A and U over G and C. This problem was posed experimentally by GAZARYAN and SHUPPE (1966, 1967). It was shown that if cells (pigeon reticulocytes) are preliminarily treated with NaF to dissociate the ribosomes from mRNA, and then the RNA is isolated from such ribosomes freed of mRNA, the nucleotide composition of the 18 S RNA approaches that of the 28 S RNA (GAZARYAN, SHUPPE, 1967) — see Table 5.

Table 5. *Nucleotide composition of high-molecular weight RNA's from animal ribosomes preliminarily freed of mRNA* (GAZARYAN, SHUPPE, 1967)

RNA Component	Nucleotide content, mole %				$\frac{G+C}{A+U}$	$\frac{Pu}{Py}$	$\frac{G+U}{A+C}$
	G	A	C	U			
28 S	36.3	16.2	30.9	16.6	2.04	1.10	1.12
18 S	34.5	17.4	30.1	18.0	1.85	1.08	1.11

On the whole, as can be seen, high-molecular weight ribosomal RNA's of vertebrate animals are characterized by a substantial predominance of G and C over A and U $[(G+C):(A+U) \simeq 2]$, a very small predominance of purine nucleotides over the pyrimidines $(Pu:Py \simeq 1.1)$, and approximate equality of the sums $G+U$ and $A+C$ $[(G+U):(A+C) \simeq 1-1.1]$.

In *invertebrate animals* and *Protozoa*, the composition of the ribosomal RNA may evidently differ substantially from the composition of vertebrate RNA and may vary rather greatly in representatives of different taxonomic groups (ANTONOV, BELOZERSKY, 1962; GUMILEVSKAYA, SISAKYAN, 1963; MEDNIKOV, 1965; PRESTAYKO, FISHER, 1966; RITOSSA et al., 1966; APPLEBAUM et al., 1966). Among Protozoa and insects, there are frequent cases when A and U predominate over G and C in the ribosomal RNA (GUMILEVSKAYA, SISAKYAN, 1963; PRESTAYKO, FISHER, 1966; RITOSSA et al., 1966).

As a summary, we should emphasize that in all the ribosomal RNA's studied, the two high-molecular weight components — 23—28 S and 16—18 S — evidently are very similar in nucleotide composition. In a number of cases, of course, substantial differences are noted. However, even if these differences actually reflect real differences in the nucleotide proportions in the two ribosomal RNA's, and not to an admixture of mRNA in preparations of 16—18 S, it is not the difference that surprises, but the great similarity. This may be directly related to the possible principal homology in the structural organisation of the two ribosomal subparticles, the larger (50 S or 60 S) and the smaller (30 S or 40 S).

The basic *general rules* of the nucleotide composition of all high-molecular weight ribosomal RNA's evidently are: 1. a greater or smaller predominance of the purine nucleotides over the pyrimidines — $G+A>C+U$; 2. an approximate equality of the number of 6-keto groups and the number of 6-amino groups — $G+U \simeq A+C$

(the Elson-Chargaff rule for RNA — see ELSON, CHARGAFF, 1954, 1955; CHARGAFF, 1963). (Again we should emphasize that frequently the small deviations from the latter equality observable as a slight predominance of the sum G+U over the sum of A+C, may be a consequence of the easy deamination of C, with its conversion to U, during the isolation of the RNA preparations and especially during their alkaline hydrolysis, so that this equality may be maintained even more rigorously in the native ribosomal RNA.)

b) Minor Nucleotides

In addition to the nucleotides of the four main nitrogenous bases (G, A, C, and U), the chains of high-molecular weight ribosomal RNA are characterized by the presence of a small amount of methylated bases (2-methyladenine, 6-methyladenine, 6-dimethyladenine, 1-methylguanine, 2-methylguanine, 5-methylcytosine, thymine, etc.), as well as pseudouridine (Ψ=5-ribosyluracil) and 2′-0-methylribose. This circumstance relates them to transfer RNA (tRNA) and distinguishes them from viral and messenger RNA's (mRNA). However, the fraction of methylated nucleosides and pseudouridine in the total number of nucleosides in ribosomal RNA's is substantially lower than in transfer RNA's (tRNA). Thus, in the high-molecular weight ribosomal RNA's of *E. coli*, there is one methyl group per approximately 100 to 150 nucleotide residues, while in tRNA there is an average of one per 30 to 40 nucleotides (DUBIN, GÜNALP, 1967; HAYASHI et al., 1966). In exactly the same way, in ribosomal RNA's of *E. coli*, one pseudouridine residue corresponds to 300 to 800 nucleotides, while in tRNA there is approximately one per 50 nucleotides (DUBIN, GÜNALP, 1967). It is interesting that the total proportion of methylated nucleotides in the 16 S RNA component is somewhat higher, and that of pseudouridine substantially lower than in the 23 S component. There are also qualitative differences; for example 6-methyladenine, 2-methyladenine, and 2′-0-methylribose are characteristic constituents of the 23 S RNA of *E. coli*, but they are not detected in the 16 S RNA; on the contrary, there is comparatively much 6-methyladenine in the 16 S RNA and none in the 23 S RNA. On the whole, calculation shows that the chain of 23 S RNA of *E. coli* contains about 10 pseudouridine residues and about 25 methyl groups, including two 2-methyladenine residues, four to five 6-methyladenine residues, one 1-methylguanine residue, three to four 2-methylguanine residues, three 5-methylcytosine residues, five thymine residues, and up to two methylated riboses. The chain of the 16 S RNA of *E. coli* evidently contains two pseudouridine residues and about 18 methyl groups, including two to three 6-dimethyladenine residues, about five 2-methylguanine residues, three 5-methylcytosine residues, and one thymine residue (DUBIN, GÜNALP, 1967). Evidently, the appearance of methylated nucleosides and pseudouridine at characteristic sites in the ribosomal RNA chains is due, just as in the case of tRNA, to the action of the corresponding methylating and modifying enzymes upon the synthesized polynucleotide chain. Methylation of ribosomal RNA occurs before or during the process of assembly of the ribosomal particles in the cell.

4. Nucleotide Sequence

High-Molecular Weight Ribosomal RNA. The complete nucleotide sequence is not yet known for any of the high-molecular weight ribosomal RNA's. However, some information on certain general characteristics of the sequence of high-molecular weight ribosomal RNA's is available.

First of all, data on a comparison of complete ribonuclease digests of the 23 S and 16 S components of the ribosomal RNA of bacteria shows that they contain different relative amounts of the various oligonucleotides and even qualitatively different oligonucleotides; consequently, the nucleotide sequences of the two high-molecular weight ribosomal RNA's — 23 S and 16 S — are definitely different (ARONSON, 1962, 1963; SANGER et al., 1965) despite the very similar, almost identical nucleotide composition. The conclusion of the difference in the nucleotide sequence of the two high-molecular weight ribosomal RNA's may also be drawn from data on the absence of competition between them in hybridization experiments with DNA, which indicates the absence of any great homology in their nucleotide sequences (YANKOFSKY, SPIEGELMAN, 1963; OISHI, SUEOKA, 1965). However, there may be some amount of homology, since opposite data on a certain partial competition between the 23 S and 16 S RNA's of *E. coli* or between the 28 S and 18 S RNA's of animal cells during hybridization with DNA have been reported [ATTARDI et al., 1965 (1, 2)].

It is very interesting that within closely related species of bacteria — for example, among species of the genus *Bacillus* — a far-reaching homology in the nucleotide sequences is detected between the same types of ribosomal RNA's in the different species (DUBNAU et al., 1965; DOI, IGARASHI, 1965, 1966). Moreover, there is a definite homology perhaps even between ribosomal RNA's of distantly related bacterial species — for example, *B. subtilis*, *S. marcescens*, and *E. coli* [ATTARDI et al., 1965 (1)], as well as among the ribosomal RNA's of all vertebrates [ATTARDI et al., 1965 (2)]. On this basis, it is concluded that there is a relative conservation of the nucleotide sequence of the ribosomal RNA in evolution.

It has been shown that the genome of the bacterial cell should contain several — up to ten — cistrons for each class (23 S and 16 S) of ribosomal RNA [YANKOFSKY, SPIEGELMAN, 1962, 1963; ATTARDI et al., 1965 (1)], while animal cells should contain up to several hundred ribosomal cistrons [RITOSSA, SPIEGELMAN, 1965; ATTARDI et al., 1965 (2)]. From this the question arises of whether the nucleotide sequence of all these cistrons is identical, and consequently, whether the nucleotide sequence of all the 23 S or 28 S ribosomal RNA and of all the 16 S or 18 S ribosomal RNA is identical in a given cell and in a given organism. There is not yet any answer to this question of a heterogeneity of high-molecular weight ribosomal RNA. On the basis of the available experiments on hybridization, it may be assumed that if the sequences of the RNA molecules of the same type are not identical, then they still should not differ too greatly, and should be very homologous, i.e., a class of the RNA molecules of the same type should represent a group of polynucleotide chains with very closely related sequences. There are some experimental indications of the possibility of heterogeneity of high-molecular weight ribosomal RNA's in bacteria (ARONSON, HOLOWCZYK, 1965; DOI, IGARASHI, 1966), but they are very indirect and the conclusion of heterogeneity is not yet convincing.

Indirect data indicate that the high-molecular weight ribosomal RNA's are characterized by definite specific features of the nucleotide sequence, which distinguish them from all other RNA's — viral, messenger, and transfer.

In the first place, ribosomal RNA's of different origins readily form complexes with sections of denatured DNA rich in C; this complex formation occurs on account of a complementary interaction of C-rich sections of DNA with regions of ribosomal RNA rich in G (OPARA-KUBINSKA et al., 1964). Both the 23 S component and the 16 S component participate in a complex formation of this type (the latter being somewhat more effective). Thus, ribosomal RNA's, in contrast to the other RNA's, exhibit a substantial fraction (from 5 to 20%) of such *G-rich regions* in the polynucleotide chain. From a comparison with data on the analysis of oligonucleotides of a ribonuclease digest (SANGER et al., 1965), however, one should think that these regions are in no case continuous poly-G blocks, but include substantial amounts of A and U (and very little C). But in any case, in these regions which undergo a complementary interaction with the C-rich sections of DNA, purine nucleotides predominate greatly, comprising two thirds of all the nucleotides (OPARA-KUBINSKA et al., 1964).

In the second place, in contrast to the other RNA's, various ribosomal RNA's exhibit hysteresis during their spectrophotometric titration in the region of pH 3 to 7 in solution, which cannot but indicate a definite specificity of their nucleotide sequence (COX, 1966). Most likely this hysteresis means a rearrangement of the secondary structure of RNA during acidification of the solution, e.g., transition from the classical complementary pairing A—U and G—C to the formation of short helices with A—A pairing (COX, 1962, 1966). In this case, these data are evidence of the presence of a definite proportion of special *A-rich regions* in the polynucleotide chains of ribosomal RNA's.

Finally, in the third place, we should recall the data on the nucleotide composition of ribosomal RNA's: purines always predominate over pyrimidines—$G+A>C+U$—and in addition, the Chargaff rule is observed — $G+U \simeq A+C$. The structural meaning of the Chargaff rule, however, unfortunately has not yet been disclosed, and for this reason, the rule has now almost been forgotten. The rule is characteristic precisely of the ribosomal RNA, and evidently it is responsible for the fact that the composition of the total RNA always more or less corresponds to this rule. The Chargaff rule is automatically derived if we assume that the rule of complementariness — $G=C$ and $A=U$ — is more or less strictly fulfilled in ribosomal RNA, but in addition, there is a *noncomplementary excess of purine nucleotides*. The structural meaning of the Chargaff rule for RNA may also lie in this. Once again the idea arises of the existence of specific regions of the ribosomal RNA chain, rich in purine nucleotides.

Low-Molecular Weight Ribosomal RNA. In contrast to the high-molecular weight ribosomal RNA, the complete nucleotide sequence of the 5 S ribosomal RNA of *E. coli* is known (BROWNLEE, SANGER, 1967; BROWNLEE et al., 1967). The 5 S RNA chain of *E. coli* is built, as has already been noted, of 120 nucleotide residues. The 5′-terminal chain residue is the 5′-phosphorylated uridine residue (pU), the 3′-terminal chain residue is also a uridine residue but with free 3′-hydroxyl (U_{OH}); the 3′-terminal trinucleotide sequence, as a whole, is CAU_{OH}. Thus, the terminals of the 5 S RNA are very specific and differ from the terminals of tRNA's (pG and CCA_{OH}). It is also difficult to find any amount of homology with nucleotide sequence of tRNA in any middle regions of nucleotide sequence of the 5 S RNA.

It is important to note that the chains of the 5 S RNA of *E. coli* do not at all display a great diversity in nucleotide sequence. In *E. coli* strain studied (MRE 600), there exist apparently only two main varieties of 5 S RNA chains, differing only in one, the 13th, nucleotide residue (G in the majority of chains, while a portion of the chains has U in position 13). In other parts the chains of 5 S RNA of *E. coli* have an identical nucleotide sequence.

The complete nucleotide sequence of 5 S ribosomal RNA of *E. coli* is shown below:

pUGCCUGGCGGCC$^{U}_{G}$UAGCGCGGUGGUCCCACCUGACCCCAUGCCGAAC-UCAGAAGUGAAACGCCGUAGCGCCGAUGGUAGUGUGGGGUCUCCCCA-UGCGAGAGUAGGGAACUGCCAGGCAU$_{OH}$.

5. Secondary Structure in Solution

It is known that isolated ribosomal RNA's in solution at low temperatures, at not too low ionic strength, and at pH values that do not differ too much from neutral, possess: a) substantial hypochromicity in the region of their specific absorption in the ultraviolet (i.e., the absorption of RNA in the ultraviolet is below the sum of the absorptions of its constituent nucleotides); b) optical activity substantially exceeding the optical activity of their constituent nucleotides; and c) great compactness of the molecule, revealed by hydrodynamic methods (sedimentation, viscosity). When the solutions are heated, or when the ionic strength is reduced below a critical level, or in the case of titration in the acid or alkaline direction, the hypochromicity, optical activity and hydrodynamic compactness abruptly drop, exhibiting features of a transition of the "order-disorder" type. The transition is entirely reversible and is purely conformational in nature. An analysis of all the enumerated data and their comparison with the properties of DNA led to the conclusion that there is a conformational regularity in the RNA molecules in solution, and that the concrete form of this regularity, at least in a substantial part, is a secondary structure represented as double helices, maintained by complementary interactions between nitrogenous bases (see surveys: SPIRIN, 1960, 1963).

At the same time, the experimentally detected properties of the secondary structure of RNA exhibit great differences from those of DNA. While DNA represents a continuous perfect double helix, imparting rigidity to the molecule along its entire length, for RNA this is not the case as is shown by its hydrodynamic properties. Furthermore, the RNA molecule is detected as a single continuous polynucleotide chain, and this means that any secondary structure existing in RNA should arise only on account of an interaction between sections of the same chain (while the DNA molecule is constructed from two chains, a separation of which can be achieved by its denaturation). Consequently, any helix formation within the RNA chain, in contrast to DNA, is created on account of an intrachain base-to-base interaction. Another distinction lies in the observable complete and practically instantaneous reversibility of the process of destruction ("melting") of the helices in the RNA molecules, while for DNA this process (denaturation) is partially reversible only under certain conditions of slow cooling (renaturation). Finally, the most important distinction observed in all experiments on the "melting" of the helical structures of RNA is the absence of the sharp narrow-temperature-interval transition, which is a characteristic feature of the structural transition of DNA molecules. The sharpness of the transition means a cooperative process of breakdown of a single regular structure. In the case of RNA, the transition occurs within an incomparably broader temperature interval than is required for the "melting" of a single helix of the DNA type, and most likely represents consecutive independent melting of numerous individual short helices.

All these experimental data led to the conclusion that helix formation in RNA molecules is not complete, as in DNA, but rather partial, and that the secondary structure of RNA in solution, in all probability, represents a set of relatively short double helices, arising on account of a complementary pairing between sections of the same chain (DOTY et al., 1959; SPIRIN et al., 1959; FRESCO et al., 1960; SPIRIN, 1960, 1961; BOGDANOVA et al., 1962).

In principle, pairing of two polynucleotide sections, with the formation of a double helix, can be achieved with the participation of the most varied pairs of bases lying opposite one another. However, DOTY and associates (DOTY et al., 1959; FRESCO et al., 1960) reported various evidences that actually in RNA molecules, the only (or in any case the substantially predominant) type of base pairing is the interaction of adenine with uracil and guanine with cytosine (A—U pairs and G—C pairs). Since helical regions in RNA molecules arise on account of bending of the chain and its sections on itself, in each given helical region the directions of the strands will be antiparallel (DOTY et al., 1959; FRESCO et al., 1960). Consequently, individual helical regions in RNA evidently are of a DNA-like character with A—U and G—C pairs and an antiparallel arrangement of the strands. X-ray diffraction studies have confirmed the presence in ribosomal RNA's of double stranded DNA-like helices with a predominant, if not exclusive A—U and G—C pairing type (ZUBAY, WILKINS, 1960; KLUG et al., 1961; SPENCER et al., 1962; SPENCER, POOLE, 1965). Finally, in a comprehensive spectrophotometric investigation of ribosomal RNA during the process of acid-base and temperature "melting" of their secondary structure, COX (1966) evidently obtained the most convincing data that the secondary structure of ribosomal RNA at neutral pH is determined *chiefly, if not exclusively, by A—U and G—C pairs.*

The quantitative determination of the helical content in molecules of ribosomal RNA in solution (i.e., the determination of what portion of all the nucleotides of the RNA chain is organized into helical regions), performed on the basis of the cited spectrophotometric data, gives 70% — at pH 7, ionic strength of the order of 0.1, and temperatures up to 25 °C (COX, 1966). Consequently, more than two thirds of the nucleotides are organized into helical regions, whereas less than one third of the nucleotide residues evidently are contained in single-stranded "amorphous" regions of the ribosomal RNA.

An analysis of the changes in the ultraviolet absorption spectra of high-molecular weight ribosomal RNA's during the process of acid-base titration and during the process of thermal "melting" of the helices shows that while the helical regions are characterized by equal ratios of purine and pyrimidine nucleotides (Pu=Py, G=C and A=U, just as could be expected from Watson-Crick pairing), *in the "amorphous" (nonhelical) regions, a substantial concentration of purine nucleotides is observed* (COX, 1966). In bacterial RNA, the predominance of purines over pyrimidines in the "amorphous" regions is substantially greater ($Pu/Py \simeq 2.5$) than in animal RNA ($Pu/Py \simeq 1.6$). All this, as is evident, agrees with the interpretation of the Elson-Chargaff rule presented above, and with the corresponding hypothesis of the presence of special, purine-rich sequences in ribosomal RNA's. A new point here is that these special, purine-rich regions exist outside the helices, i.e., evidently form single-stranded links between the helical regions within the whole secondary structure of ribosomal RNA in solution.

From a detailed analysis of the same spectrophotometric data, it follows that ribosomal RNA's, both bacterial and animal, contain *two discrete classes of helical regions*, differing in nucleotide composition: helices with an approximately equal ratio of the G – C and A – U pairs, and helices where the G – C pairs substantially predominate over A – U pairs (Cox, 1966). Correspondingly, the helical regions with a predominance of G – C pairs are more thermally stable. In animal ribosomal RNA (rabbit reticulocytes), helices of the second type contain an average of about 85% G – C pairs (of the total number of pairs); in the bacterial ribosomal RNA *(E. coli)*, the percentage of G – C pairs in the thermally stable helices is lower — about 67%.

As for the question of the size (length) of the helical regions contained in RNA, the work of Doty and associates (Doty et al., 1959) indicated that the helical regions in any case cannot be very long. According to these data, especially from the data on the widths of thermal transition and the mean melting points, the secondary structure of RNA evidently represents a set of rather various short helical regions, distributed over the entire RNA particle. A more detailed analysis of this question was undertaken by Fresco et al. (1960), who deduced that the minimum size of the helices existing in RNA is evidently four to six nucleotide pairs (about half of a turn of the double helix), but the bulk of the helices evidently are longer, and on the average the size of the helices may comprise about one full turn (ten nucleotide pairs). This conclusion was entirely confirmed especially for ribosomal RNA's by Cox (1966), who demonstrated that the number of nucleotide pairs forming the helical region is small and lies mainly between 4 and 17.

Hence, the polynucleotide chain of high-molecular weight ribosomal RNA in solution forms numerous *short* double helical regions on account of pairing of sections of the chain, with an *antiparallel* direction of the interacting sections. A number of general considerations and indirect experimental data force one to believe that the short helical regions with an antiparallel direction of the strands can and should arise in RNA chiefly on account of pairing of *adjacent* (neighboring) sections of the polyribonucleotide chain (Fresco et al., 1960; Spirin, 1960; Bogdanova et al., 1962). A scheme of a fragment of such a structure is given in Fig. 14a.

The existence of numerous short helical regions, involving *adjacent* sections of the chain, automatically presupposes that the helical regions on the whole form a set of "hairpins", protruding in all directions from the connecting backbone of the polynucleotide chain (Fig. 14a). Evidently in the case of low ionic strength (0.1 to 0.01), when the electrostatic repulsion among the phosphates can be still sufficiently pronounced, the helical regions should have a tendency for mutual repulsion, and thereby the entire molecule should tend to be stretched out in a direction perpendicular to the axes of the helices; thus, a strand-like or rod-like structure of high-molecular weight RNA in solution can arise. This structure will thus be maintained solely by forces of electrostatic repulsion which stretch out the particle, in the presence of a covalent backbone which successively links the helical regions to one another. If a significant rôle is attributed also to the metal bridges and non-specific hydrogen bonds within the RNA macromolecule, then the possibility of some attraction or association between the parallelly arranged neighboring helical regions themselves arises, forming a structure similar to the variation shown in Fig. 14b. The possibility is not excluded that in the case of parallel packing of the helical regions within the RNA macromolecule, zones of crystalline packing may even arise.

Thus, summarizing the considerations cited with respect to the principles of the mutual arrangement of the helices in high-polymer RNA's in solutions at room temperature and moderate ionic strength, it may be stated that the short "hairpin-like" double helix regions in the RNA molecules should possess a tendency to be arranged

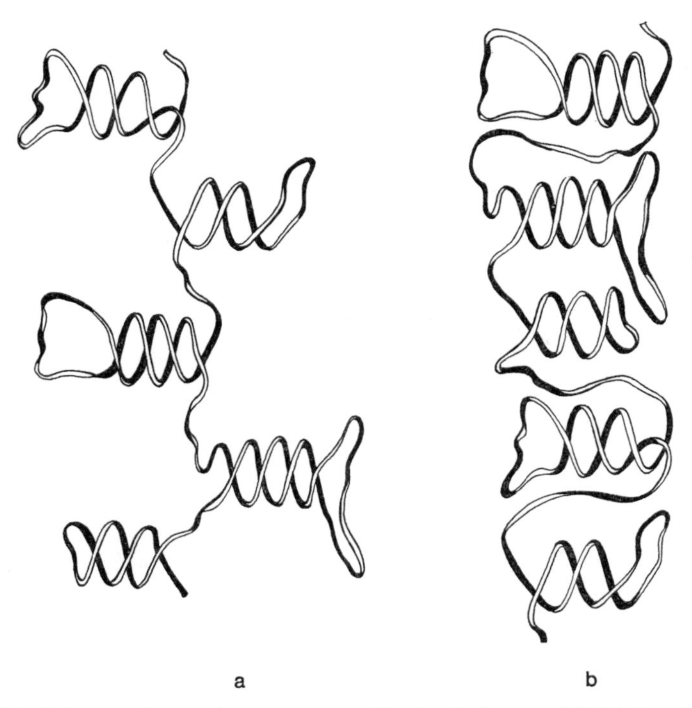

Fig. 14a and b. Scheme of secondary structure of isolated ribosomal RNA in solution (fragment of molecule). Two possible variations of mutual arrangement of the helical regions: a) in the absence of positive interactions between the helical regions; b) in the presence of positive interactions (mutual packing) between the helical regions (Bogdanova et al., 1962; Spirin, 1963)

primarily perpendicular to the long axis of the entire molecule, alternating with non-helical single-stranded regions. The "piling" of these helical regions and the random regions alternating with them forms a rod-like particle.

Independent experimental confirmations of the *rod-like* or *strand-like* shape of the molecules of high-molecular weight ribosomal RNA's in solutions of moderate ionic strength were obtained from light scattering measurements (Timasheff et al., 1958; Kronman et al., 1960), and from electron microscopy (Kisselev et al., 1961; Bogdanova et al., 1962). The diameter of the rods or strands has been estimated of about 30 Å, which is in good agreement with the average *length* of a helical region with a dimension of about one turn. The general size of the "rods" observed in an electron microscope agrees with the proposed model of the rod-like structure of high-molecular weight RNA in solution.

On the whole, in a consideration of the secondary (and tertiary) structure of ribosomal RNA in solution, it should be recalled that this is not a stable structure, pre-existing in the ribosome and preserved during deproteinization. A number of methods of deproteinization (for example, hot deproteinization with phenol) are *known* to

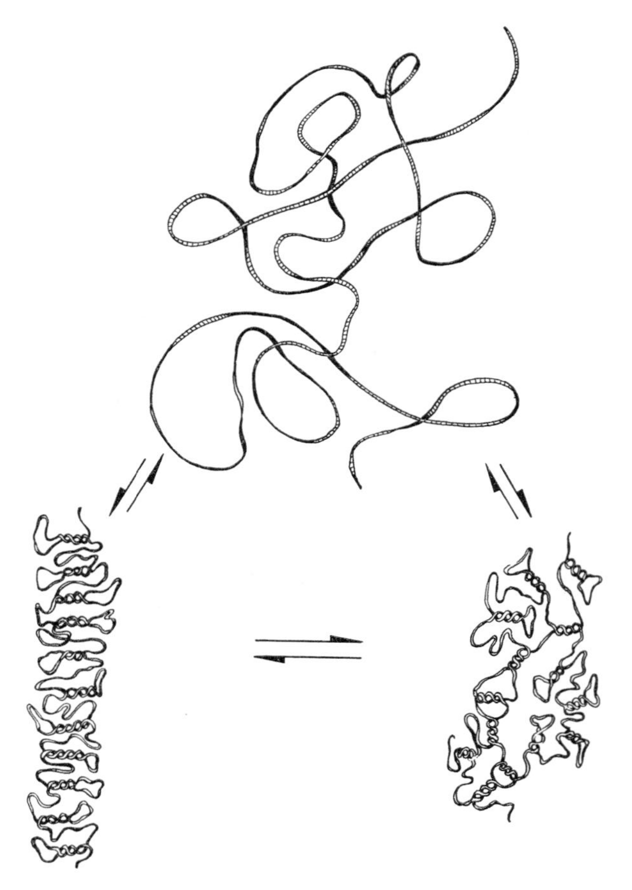

Fig. 15. Scheme of conformations and conformational transitions of high-polymer RNA macromolecule in solution as a function of ionic strength, temperature, and pH ("coil — rod — extended strand"). (Spirin, 1960, 1963)

"melt" all or almost all of the secondary structures during the treatment. Nonetheless, in solutions with moderate ionic strength, high-molecular weight ribosomal RNA, isolated by any method, possesses a secondary structure, evidently with all the features characteristic of ribosomal RNA which were described here. Consequently, in a description of the secondary structure of RNA, it evidently cannot be considered as any primordially set, fixed, native structure. On the contrary, this structure is a typically equilibrium reversible structure. Under concrete conditions of a solution, the RNA chain spontaneously acquires this conformation which under these conditions is the result of an equilibrium of different molecular forces, and corresponds to the minimum free energy. It may be thought that RNA acquires its own most stable conformation as a result of lateral mobility, which makes it possible to explore and find the most advantageous of the competitive conformations. The complete and practically instantaneous reversibility of all the changes in the secondary structure and general conformation, including the restoration of all the previous characteristics — the helix content, particle size and shape — after complete "melting" of the helices and thermal unfolding of the chain (BOGDANOVA et al., 1962) is in full agreement with these concepts. Evidently each RNA molecule under given conditions of the solution will spontaneously take up its own concrete "pattern" of the secondary structure, the uniqueness of which will be dictated by the specific nucleotide sequence of the RNA.

In the case of a significant deviation of the conditions of the solution from those described (low temperature, neutral pH, moderate ionic strength), the conformation of the ribosomal RNA molecules, just as of all the other RNA's in solution, undergoes reversible conformational transformations (see survey: SPIRIN, 1960, 1963). Depending upon the conditions of ionic strength, temperature, and other factors, RNA macromolecules can exist in different conformations. At moderate ionic strength and room temperature, they represent *compact rods*, made up of short helical regions, alternating with random regions. When the ionic strength is increased, they become less asymmetrical and pass into a state more resembling *compact coils*. When the temperature of the solutions is increased or upon transferring them into salt-free solutions, as well as after acidification, alkalinization, or under the action of formamide, dimethyl sulfoxide, etc., the macromolecules pass into the state of more or less *extended threads* or *loose random coils*. All these extreme conformations are interrelated by reversible transitions. The indicated representations are given schematically in Fig. 15.

6. Secondary Structure within the Ribosome

The most important question in the study of the structure of RNA in the isolated free state in solution is the question of the relationship of the detected structure to the biological functions of the molecule and to the native structure possessed by this RNA in functioning living systems. For ribosomal RNA, the problem is especially complex, since in the cell it generally never exists by itself, but is always organized in a definite way with protein into specific ribonucleoprotein particles. The main question for ribosomal RNA is what structure high-polymer RNA possesses within the ribosomal particle, and to what degree the principles and features of the molecular organization of free RNA in solution discussed are pertinent to RNA within the ribosomes.

The present available data indicate that the general nature of the helical regions and the helix content of ribosomal RNA's, in all probability, are approximately the same in free RNA and RNA contained within the ribosomes.

These are primarily data on the study of the UV hypochromicity of the ribosomes, or more strictly, of the RNA within the ribosomes. The first experiments were conducted by HALL and DOTY (HALL, DOTY, 1959), who demonstrated that RNA within the ribosomes is hypochromic, i.e., may contain helical regions, maintained by complementary interactions, and that when solutions of the ribosomes are heated, these helical regions "melt" similarly to those in free RNA. In 1960, a whole series of authors (ZUBAY, WILKINS, 1960; SCHLESSINGER, 1960; BONHOEFFER, SCHACHMAN, 1960) attempted a quantitative comparison of the hypochromicity of RNA within the ribosomes and of isolated RNA. It was found that the extent of hypochromicity of RNA within the ribosomes is identical with that of free RNA in solution with high ionic strength. These experiments indicated that the number of helical regions and portion of nucleotides included in helices approximately coincide in free RNA in the coiled state and in RNA within the ribosomes. It may be concluded that the data on the secondary structure of RNA in solution are to some degree valid for RNA within the ribosomes as well, and that in the ribosomes the helix content and basic features of the secondary structure of RNA, which it exhibits in the free form, are not perturbed or disrupted by its interaction with the ribosomal protein.

Another group of data is represented by X-ray diffraction studies of the ribosomes (ZUBAY, WILKINS, 1960; KLUG et al., 1961). In these investigations it was shown that the diffraction pattern of the ribosomes themselves practically does not differ from the pattern obtained by superposing the diffraction pattern of isolated RNA on the diffraction pattern of free ribosomal protein. An analysis of the diffraction patterns of the ribosomes indicated that RNA in the ribosomes is evidently characterized by helical regions of the same nature and structure as free RNA in solution; these regions evidently are double-standed (double helical); their chains are antiparallel, and the basic type of base pairing in them is A—U and G—C pairs.

Finally, the third group of data pertains to the study of the optical rotatory dispersion of the ribosomes in comparison with isolated ribosomal RNA in solution (BLAKE, PEACOCKE, 1965; MCPHIE, GRATZER, 1966; SARKAR et al., 1967). It is known that RNA in the region of its intrinsic absorption in the ultraviolet exhibits a very characteristic Cotton effect, *very sensitive to conformational changes* (changes in the secondary structure), with a peak near 280 mμ and a trough near 250 mμ. It was found that the ribosomes give *qualitatively and quantitatively* exactly the same Cotton effect in this spectral region as the RNA isolated from them. On the whole, the optical rotatory dispersion profile of the ribosomes in the entire region from 220 to 350 mμ is evidently identical with the simple sum of the optical rotatory dispersion profiles of the free ribosomal RNA and ribosomal protein. The indicated observations were made on bacterial, yeast, and animal ribosomes. Consequently, even such a fine and sensitive method as measurement of optical rotatory dispersion did not reveal any differences of the secondary structure of ribosomal RNA within the ribosomes and in the isolated form at moderate ionic strength.

Thus, the various methodological approaches to the study of the internal structure of the ribosomes lead to the same conclusion: the secondary structure of RNA within RNP particles (ribosomes) seems to be quite similar to the secondary structure that is exhibited by free high-polymer RNA in solutions with moderate salt concentration, and hence the conclusions cited above (Section II, 5) on the nature of the latter may be valid for a description of the structure of RNA within the native functioning particles[1].

[1] There are opposite indications in literature as well. FURANO, BRADLEY and CHILDERS (FURANO, BRADLEY and CHILDERS, 1966) studied the absorption spectre change of acridine orange as a result of its binding with ribosomes (with RNA in ribosomes) and with isolated ribosomal RNA. They found that nearly 90% of RNA phosphates in the ribosomes are free for rapid binding of the dye. The change in the acridine orange absorption spectre during its

The concept of preservation of the principal features of the secondary structure of free RNA within the ribosomes, i.e., despite its interaction with a substantial amount of protein, suggests a consideration of possible ways of mutual arrangement of the RNA and protein molecules in the ribosome. One of the ways is that which can be observed in "spherical" viruses: a definite "segregation" of the ribosomal RNA from protein, with the creation of a core of RNA, covered with a shell of protein molecules. This way presupposes a minimal interaction between RNA and the protein molecules. However, electron microscopy of the ribosomes, and in particular, using positive staining of RNA with uranyl acetate, definitely gives evidence against the presence of a protein shell around RNA or in general of any concentration of RNA in the center of the ribosomal particles; the RNA and protein *seem to be uniformly "mixed"* and interpenetrate each other through the whole ribosome (Huxley, Zubay, 1960). Furthermore, experiments on the effects of exogenous ribonuclease upon the ribosomes show that despite the preservation of apparent formal integrity of the particles, high-molecular weight ribosomal RNA within the ribosomes is subjected to rather substantial degradation by the exogenous enzyme (Shakulov et al., 1962; see also the report of Santer, 1963). This means that definite regions of the ribosomal RNA are open for contact with the surrounding macromolecules, i.e., posses a surface arrangement or are accessible from the surface.

On the basis of data on the reversible structural transformations of the ribosomes, primarily on their unfolding into ribonucleoprotein strands (see below, Section IV, 2; Spirin et al., 1963; Spirin, 1963), the following may be taken as the most probable hypothesis: 1. the chain of high-molecular weight ribosomal RNA forms a *secondary structure*, described above for the case of RNA in solution, with numerous short helical regions, successively connected through intervening single-stranded regions (Fig. 14) into a *flexible "rod" or strand;* 2. protein molecules interact with RNA chiefly *at the non-helical "amorphous" regions* of the chain connecting the helices, so that the secondary structure is not perturbed, and the *ribonucleoprotein strand* is formed; 3. the ribonucleoprotein strand, the structural frame (skeleton) of which is an RNA rod, in turn is *folded in a definite compact way*, forming the ribosomal particle (correspondingly, depending upon the RNA, 50 S or 30 S). This mode of mutual organiza-

---

titration with ribosomes differed from the change of the spectre during titration of the dye with isolated ribosomal RNA; this difference fits into the concept of a very low content of double helices in RNA within the ribosome in contrast to the isolated RNA's. However, no control experiments were made in the considered work which would have shown that the ribosomes themselves do not degrade and do not change during dye binding. In recent investigations by Permogorov and Sladkova (Permogorov and Sladkova, 1968), the cited work was subjected to a thorough verification. With this end in view, a parallel study was made of the changes in the absorption spectre of acridine orange, of the changes in the optical rotatory dispersion in the region of the absorption of acridine orange, and of the changes of sedimentation properties of ribosomes during titration of acridine orange with ribosomes. The results of the examination showed that when the ribosome-to-dye ratio is low (when there is an excess of dye) the ribosomes are changed or broken down and the number of phosphates accessible to the dye becomes considerably higher than in the intact ribosomes, reaching 90%. As for the undisturbed ribosomes, the number of phosphates accessible to the dye is only about 20% of the number of accessible phosphates groups in isolated ribosomal RNA. It is thus apparent that the interpretation of the results of spectrophotometric titration in the work of Furano et al., is erroneous and cannot serve as any basis for the assessment of the secondary structure of RNA within ribosomes.

tion of RNA and protein within the ribosome permits an explanation both of the preservation of the "free" secondary structure of RNA within the ribosome and of the apparent intermixing and interpenetration of RNA and protein through the entire volume of the ribosomal particle.

III. Ribosomal Proteins

1. Definition

Each ribosome in the cell or in the isolated state contains *many* protein molecules. However, the degree of binding and the strength of retention of various proteins in the ribosome may be quite different. Moreover, certain ribosomal proteins may not be bound at all firmly and constantly within the ribosome, but exist in a dynamic equilibrium with the exogenous free protein pool in solution and thus are readily exchangeable for analogous exogenous proteins and are readily removed when the ribosomes are washed. On the other hand, proteins that do not have any direct relationship to the functioning of the ribosome can evidently be sorbed upon it. Since ribosomes must be studied chiefly in the isolated state, it is evident that the use of various methods of isolation and purification should lead to the isolation of preparations with different protein content. This is actually the case. As a result, great difficulties arise in the determination of what should be considered truly ribosomal proteins.

In this analysis, we shall consider true ribosomal proteins or *structural ribosomal proteins* to be only those proteins of the ribosomes that are not removed from the particles during thorough purification of the preparations. Thorough purification means repeated resedimentation in a preparative ultracentrifuge, passage through a sucrose gradient, or best of all washing with 0.5 M or 1 M NH_4Cl or KCl — all in the presence of Mg^{++} ions (10^{-4} M to 10^{-2} M). The corresponding ribosome preparations will be denoted as *thoroughly purified ribosomes.* The term "structural proteins" should be taken quite conditionally in this case, in the following sense: each of these proteins may possess no independent functional significance, but within the ribosomes they participate in the formation of *structures* possessing both the functions of specific *binding* of other components of the protein-synthesizing apparatus and *catalytic* functions.

In the *crude ribosomes,* isolated by the usual methods of one, two, or three sedimentations in an ultracentrifuge in the presence of 0.005 M to 0.02 M Mg^{++}, in addition to the indicated "structural ribosomal proteins", there may be: a) loosely bound proteins, which also play a definite rôle in the structure of the ribosome and in the association with other components of the protein-synthesizing system; these, for example, include the so-called "initiation factors" (see Part II, "Functioning of the Ribosome", Section I); b) proteins which, combined with the ribosome (and only combined with it), are responsible for certain catalytic functions of the particle as a whole; these include, for example, the "transfer factor" G (see Part II, "Functioning of the Ribosome", Section I); c) enzymes whose activity is provided for by their intrinsic structure, and is not induced by the ribosome (this means that they are entirely active in the free state in solution). In all probability, all these loosely bound proteins are always present in the cellular sap in a greater or lesser excess. Therefore, although thoroughly purified ribosomes do not contain them, and hence cannot themselves synthesize protein in the presence of mRNA and aminoacyl-tRNA, they again may become full-valued biologically active ribosomes when the cellular sap is added. It might be thought that in the cell or in a full-valued protein-synthesizing cell-free system, the indicated labilely bound proteins exist in dynamic equilibrium with the exogenous free protein pool. These proteins may be denoted as *conditionally ribosomal proteins*.

In addition to these proteins, preparations of crude ribosomes may contain certain amounts of enzymes or other protein components sorbed by the ribosomes during the process of isolation and possessing no direct relationship either to the structure or to the functioning of the ribosomes. Evidently such *extraneous proteins* should include, in particular, most of the enzymes detected in association with the ribosomal particles: deoxyribonuclease, polynucleotide phosphorylase, ATPase, phosphatases, peptidases, dehydrogenases, etc. (see the survey: PETERMANN, 1964).

The question of ribonuclease requires a special consideration. Ribonuclease activity is characteristic of most ribosomal preparations, and therefore ribonuclease is frequently considered even as a necessary component, intrinsic to the ribosome (see survey: PETERMANN, 1964). The ribonuclease contained in the ribosomes is usually in an inactive, latent state. Activation of the ribonuclease occurs in a solution of comparatively high ionic strengths (0.1 and above), or when Mg^{++} ions are removed, or under the action of urea, as well as under the action of a whole series of other agents that damage the ribosome. Such activation rapidly leads to degradation of the ribosomal RNA and breakdown of the particles, i.e., to "self-digestion" of the ribosome preparation. It has been shown on ribosomes of *E. coli* that the enzyme — ribonuclease I — is specifically localized in the 30 S subparticle (ELSON, TAL, 1959; WALLER, 1964). However, it was found that the enzyme can be washed off from the ribosomes of *E. coli* with 0.5 M to 1 M NH_4Cl (STANLEY, BOCK, 1962, unpublished; SPIRIN et al., 1963; SALAS et al., 1965), and in this case their functions in the protein-synthesizing system *in vitro* are not impaired. Later species of bacteria were found, the ribosomes of which do not have any latent ribonuclease at all — for example, *Bacillus megatherium* (WALLER, 1964), *Pseudomonas fluorescens* (CAMMACK, WADE, 1965), etc. Even for *E. coli*, mutants containing no ribonuclease I have been found in nature (CAMMACK, WADE, 1965) and have been selected artificially [GESTELAND, 1966 (1)]. Finally, it has been shown (NEU, HEPPEL, 1964) that in normal cells of *E. coli* the bulk of the enzyme is present before breakdown of the cells not in the ribosomes, but between the cell wall and the cell membrane; breakdown of the cell leads to a release of ribonuclease I from the surface layer into the solution, after which it is sorbed on the ribosomes, specifically associating with the 30 S subparticle with a transition to the inactive (latent) state. Thus, evidently even an enzyme so characteristic of the ribosomes as ribonuclease is found to be more likely an extraneous protein for the ribosome. Nevertheless, the ability of the enzyme to associate specifically with the ribosome (precisely with the 30 S subparticle) and its specific inactivation on the ribosome requires its own explanation.

In the exposition to follow, pertaining to the structure of the ribosomal proteins, we shall discuss neither the conditionally ribosomal proteins nor, all the more the extraneous proteins sorbed on the ribosomes, but shall limit ourselves to a consideration only of the ribosomal proteins that remain in the ribosomes after their thorough purification and washing, i.e. — *only the structural ribosomal proteins.*

2. Number of Protein Molecules per Ribosome and Their Molecular Weight Characteristics

a) Bacterial 70 S Ribosomes

In purified bacterial ribosomes, protein comprises about one third of the dry weight of the ribosome. This means that the total "molecular weight" corresponding to all the protein of the 70 S ribosome is about 10^6 daltons.

According to the chemical determination made on the basis of the analysis of the numbers of the N-terminal groups, the average length of the polypeptide chains of the total ribosomal protein of *E. coli* was found to be equal to approximately 230 amino acid residues which corresponds to a molecular weight of about 25000 (Waller, Harris, 1961). The similar value of 26000 of the average molecular weight of the polypeptide chains of the total ribosomal protein of *E. coli* was obtained from data on sedimentation and diffusion in conditions of dissociation and unfolding of the chains in 8 M urea ($S^0_{20,w} = 1.1 - 1.35$ S) (Möller, Chrambach, 1967).

However, the above mentioned data concerning the chemical determination of the average length of polypeptide chains, taking into account the number of free terminal N-groups, could not serve as a sound basis for judging the average size of the chains of ribosomal proteins, inasmuch as it was not known whether all the N-terminal groups of the total ribosomal protein were free; if part of the polypeptide chains of ribosomal protein have blocked N-terminal groups, then apparently such a determination must give an overestimated value of the average size of polypeptide chains. And, further, the sedimentation method of determining the weight-average molecular weight of ribosomal protein also presents a danger of overestimating the values of molecular weight, in this case due to the probable presence in the solution of a portion of polypeptide chains in the form of aggregates. Indeed, the determination of weight-average molecular weight of polypeptide chains of ribosomal protein of *E. coli* by the approach to sedimentation equilibrium method of Archibald (modified by Trautman) or by the sedimentation equilibrium method (modified by Yphantis) when all special precautions against the presence of aggregates were taken, gave a value even as low as 14,000 to 15,000 (Ivanov, 1967, unpublished; Moore et al., 1968). Moreover, an examination of N-terminal groups in individual fractions of ribosomal protein showed that some of them do not manifest free N-terminal groups (Traut et al., 1967; Moore et al., 1968). At present the value of about 20,000 is usually accepted as a more real figure for average molecular weight of the ribosomal proteins; various estimates give a figure of about 60 for the number of polypeptide chains (protein molecules) per ribosome of *E. coli;* of this number, about 40 must be attributed to the 50 S particle and about 20 to 23 to the 30 S subparticle.

Ribosomal protein is not homogeneous *but very heterogeneous* with respect to molecular weight. This is best of all demonstrated by the data obtained by H. Wittmann's laboratory (KALTSCHMIDT et al., 1967): the ribosomal protein of the 30 S subparticle was separated into a large number of fractions, many of which represented individual proteins, and the molecular weights of the majority of these individual proteins were directly determined by the sedimentation equilibrium method. The determination resulted in the following values of molecular weights for 15 individual fractions of ribosomal protein of the 30 S subparticle: 9,000, 11,000, 12,500, 13,500, 15,000, 18,000, 19,500, 20,000, 22,000, 26,500, 28,000, 35,000, 36,000, 41,000, 41,000. Thus, the range of the variety of sizes of polypeptide chains of ribosomal protein is very great, from polypeptides consisting of about 100 amino acid residues (molecular weight about 10,000) to large protein chains of 300 to 400 amino acid residues (molecular weight 30,000 to 40,000). It is important that each polypeptide chain is represented in the ribosome, at any rate in the 30 S subparticle, apparently by a single copy (MOORE et al., 1968).

b) Animal 80 S Ribosomes

The 80 S animal ribosomes have a far greater relative and absolute protein content than bacterial ribosomes: the protein in them takes up nearly half the dry weight and the "molecular weight" a falling on the total protein of the 80 S particle is $2-2.5 \times 10^6$ daltons. According to the early determination of the molecular weight of protein of the ox liver ribosome by the method of equilibrium ultracentrifugation, the weight-average molecular weight of polypeptide chains is about 25,000 (CURRY, HERSH, 1962). However, a later conducted determination of number-average molecular weight by terminal group analysis according to the data obtained on total protein of the rabbit reticulocyte ribosome gives considerably lower values, only about 12,000 to 14,000 (MATHIAS, WILLIAMSON, 1964). A value approaching this, about 16,000 or 17,000, can be derived for the number-average molecular weight of the rat liver ribosomal protein as an average of the molecular weights of the fractions of this protein, analyzed by the method of equilibrium ultracentrifugation (HAMILTON, RUTH, 1967). Consequently, the number of polypeptide chains of the ribosomal protein per 80 S ribosome of animal origin is evidently somewhere around 150, and of these about 100 can be attributed to the 60 S subparticle and about 50 to the 40 S subparticle.

In a study of the molecular weights of individual fractions of the ribosomal protein of the 60 S subparticle of rat liver ribosomes, values of the weight-average molecular weight from 10,000 to 30,000 have been found (HAMILTON, RUTH, 1967). One of the electrophoretically homogeneous fractions, comprising about 20% of the total protein of the 60 S subparticle, gave a molecular weight of about 29,000; another electrophoretically homogeneous fraction, encompassing somewhat more than 10% of the total protein, gave a molecular weight of about 15,000. More than 20% of the total protein of the 60 S subparticle was distributed in the fraction with a molecular weight of about 10,000 to 11,000. Consequently, the protein of animal ribosomes, like the protein of bacterial ribosomes, is *very heterogeneous* with respect to molecular weight; a substantial portion of the total number of protein molecules consists of relatively low-molecular weight proteins (polypeptides).

3. Amino Acid Composition and Sequence

As a rule, the amino acid composition of the total ribosomal protein (Ts'o et al. 1958; Crampton, Petermann, 1959; Spahr, 1962; Mathias, Williamson, 1964) is characterized by a comparatively high content of basic amino acids — arginine, lysine and histidine — which on the whole imparts a basic character to it. The protein contains very little cysteine and probably possesses no disulfide bridges at all. In other respects, the set of amino acids seems to be full-valued: there are all the amino acids characteristic of the usual globular proteins, including tryptophan. The content of nonpolar amino acids is at the level characteristic of globular proteins, rich in nonpolar amino acids. Table 6 presents as an example the composition of the total ribosomal protein of *E. coli* according to the data of Spahr (Spahr, 1962).

A very great similarity of the amino acid composition of the total protein of the two ribosomal subparticles is noted. Thus, the total proteins of the 50 S and 30 S subparticles of the ribosomes of *E. coli* practically do not differ in the content of most amino acids, and small, statistically significant differences are detected only for five amino acids: the protein of the 50 S subparticle contains slightly more alanine and threonine and slightly less arginine, histidine and glutamic acid than the protein of the 30 S subparticle (Spahr, 1962). In exactly the same way, the total protein of the 50 S subparticle and that of the 30 S subparticle are very similar in N-terminal amino acids: in both about 46 to 49% of the terminal groups are represented by methionine, about 36 to 40% by alanine, about 10 to 12% by serine, 1 to 3% by threonine, and 1 to 2% by glutamic and aspartic acids (Waller, Harris, 1961).

Any preparation of total ribosomal protein is not homogeneous, but can be separated into a greater or smaller number of fractions, differing in electrophoretic mobility in gels (Setterfield et al., 1960; Waller, Harris, 1961; Waller, 1963, 1964; Leboy et al., 1964; Traut, 1966; Traub et al., 1966; Möller, Chrambach, 1967; Low, Wool, 1967; Hamilton, Ruth, 1967) or in binding on cation exchange columns (Waller, Harris, 1961; Waller, 1963, 1964; Spitnik-Elson, 1964; Möller, Widdowson, 1967). These differences, at least to certain degree, unquestionably reflect the heterogeneity of the total ribosomal protein with respect to basicity, i.e., with respect to amino acid composition. Thus, it can be concluded that the structural protein of the ribosomes represents a set of polypeptide chains of various basicities and different amino acid compositions.

This heterogeneity is great. The total protein of the ribosomes of *E. coli* can be resolved into 20 to 35 and even more separate bands by electrophoresis in gels; most of the bands characteristic of the protein of the 50 S subparticles (up to 20 and more bands) do not coincide with the bands characteristic of the protein of the 30 S subparticles (up to 12 to 15 bands) (Waller, 1964; Leboy et al., 1964; Traut, 1966; Traub et al., 1966). Consequently, a) the structural protein of the 50 S subparticles differs from the protein of the 30 S subparticles, and hence each of the two subparticles is characterized by its own *specific set* of polypeptide chains; b) the set of polypeptide chains in each ribosomal subparticle is *quite various*.

Despite the exceptional heterogeneity of the total ribosomal protein or the protein of the individual ribosomal subparticles, the entire set of polypeptide chains can be broken down into several groups according to their basicity. Thus, on the grounds

Table 6. *Amino acid composition of total ribosomal protein of E. coli (according to data of* SPAHR, *1962)*

Amino acid residues	Content, in moles per 100 moles of amino acids (mole %)		in grams per 100 g of protein (% by weight)
Nonpolar			
Glycine	8.2		4.2
Alanine	11.0		7.0
Valine	9.6	27.9	8.6
Leucine	7.4		7.5
Isoleucine	5.5		5.6
Phenylalanine	3.0		4.0
Methionine	2.4		2.8
Proline	3.7		3.2
Cysteine	0.5		0.5
Tryptophan	0.7		1.2
Tyrosine	1.8		2.6
Polar, nonionized			
Serine	4.4	16.7	3.4
Threonine	5.2		4.7
Asparagine + glutamine	7.1		7.8
Acidic			
Aspartic acid + glutamic acid	11.3		12.5
Basic			
Arginine	7.2	18.1	10.3
Lysine	9.0		10.4
Histidine	1.9		2.4
Excess of basic groups over acidic groups (arg + lys + his) − (asp + glu)	6.8		—

of chromatographic separation data on carboxymethylcellulose, the total ribosomal protein of *E. coli* can be broken down into four groups: 1. a small group of acidic proteins; 2. a large group of weakly basic proteins; 3. a large group of moderately basic proteins; 4. a small group of strongly basic proteins (MÖLLER, WIDDOWSON, 1967; GARBER and BARULINA, 1968). These four groups are present both in the 50 S and in the 30 S subparticles. The indicated groups differ in amino acid composition and especially in content of acidic and basic amino acids. The small group of acidic proteins (encompassing about 7—8% of the total ribosomal protein) is characterized by an especially high content of alanine (about 18%, in moles per 100 moles of amino acids), an increased content of glutamic acid, valine and cysteine, and a low content of arginine (only about 3%) and tryptophan. The groups of basic proteins differ chiefly in the content of aspartic acid, glutamic acid, glycine, histidine, lysine and arginine; they are very similar in the content of all the hydrophobic and aromatic amino acids (MÖLLER, WIDDOWSON, 1967).

However, although such a more or less distinct subdivision of the total ribosomal particle of *E. coli* into four groups, differing in basicity, is objectively observed, not one of the groups is chemically homogeneous. Less complicated is the group of acidic proteins, about two thirds of which (5% of the total ribosomal protein) are represented by identical, or to be more correct, related in *composition and sequence*, polypeptide chains with a molecular weight of about 20,000 to 25,000; it is these chains that are characterized by a very high alanine content (20%), a very low content of histidine (0.8%) and arginine (1.8%), and a total absence of tryptophan (Möller, Widdowson, 1967). Each of the three groups of basic proteins represents very complex mixtures of chemically different polypeptide chains; this follows both from the multitude of electrophoretic bands exhibited by each group, and from the great complexity of their peptide maps (Möller, Widdowson, 1967).

In a special investigation of ribosomal protein of the 30 S subparticles of *E. coli* it was shown on the basis of chromotography data on carboxymethylcellulose and then on Sephadex, that it could be separated into 20 individual electrophoretically homogeneous fractions which can be characterized as individual homogeneous proteins (Traut et al., 1967; Moore et al., 1968). A study was made of their amino acid composition, N-terminal groups, peptide maps of their tryptic hydrolyzates and finally of their molecular weight and the number of molecules of each species per the ribosomal particle. In this way, with the ribosomal protein of the 30 S subparticle of *E. coli* as an example, the following conclusions can be drawn; a) *all* the 20 polypeptide chains within this subparticle are *different* and each species of polypeptide chain is represented in the 30 S subparticle by only one copy; b) there is *no similarity* or homology between any of them in their amino acid sequence and they do not group themselves in any "families" of related chains; c) all the 30 S particles in the cell have *the same set* of polypeptide chains, i.e., the population of the 30 S subparticles is homogeneous in protein composition (Moore et al., 1968). It may be suggested that in a whole ribosome all its 65 protein molecules (polypeptide chains) are different and that each plays its own specific, structural or functional role within the ribosome.

The ribosomal proteins are evidently species-specific with respect to amino acid composition and sequence. In any case, electrophoresis in gels of ribosomal proteins of different species of bacteria — *E. coli*, *M. lysodeikticus*, *B. cereus*, and *B. megatherium* — shows a quite characteristic band distribution profile for each of them (Waller, 1964). On the contrary, different strains of the same species of bacteria *(E. coli)* as a rule give identical band distribution profiles during electrophoresis. However, here too, differences in single proteins may be found in some cases; for example, in all the strains of K 12 there is a basic ribosomal protein differing in electrophoretic mobility from the corresponding protein in other strains (Leboy et al., 1964). Substantial deviations in the total amino acid composition of the ribosomal protein, chiefly in the direction of a significant increase in the acidity of most of the polypeptide chains, are observed as a result of the transition to halophilism, as has been demonstrated on halophilic (living in the presence of high salt concentrations) bacterium, *Halobacterium cutirubrum* (Bayley, 1966).

The proteins of animal ribosomes naturally give their own characteristic electrophoretic band distribution profile, quite different from the profile of the protein of bacterial ribosomes. For the ribosomes of mammals (rat, rabbit), it has been found that all the polypeptide chains seem to be basic (patently acidic proteins are not detected) (Low, Wool, 1967). Ribosomes from different tissues of the same species (rat), and even ribosomes of another, true, relatively close species (rabbit) give qualitatively quite similar band distribution profiles.

4. Conformation

Little is known about the conformation of the polypeptide chains of the ribosomal protein. A substantial difficulty in its study is that the separation of proteins from the ribosome, as a rule, is accompanied by a greater or lesser denaturation and aggregation of it. That is why all the electrophoretic, sedimentation, and other studies of isolated ribosomal protein are conducted, as a rule, in the presence of 8 M urea, where protein is represented in the form of disaggregated unfolded polypeptide chains. It is clear that a study of protein in such a solvent or in the aggregated state can contribute little to a clarification of its native conformation. Moreover, it is probable that the ribosomal proteins possess a quite stable native conformation *only within the ribosome*, and *any* removal of them from the ribosome either substantially destabilizes their native conformation or even automatically leads to denaturation.

Attempts to investigate the conformation of the ribosomal protein within the ribosome have been undertaken by means of a study of the optical rotatory dispersion of the ribosomes in the region of the intrinsic ultraviolet absorption of the peptide bond — 200 to 230 mμ (McPhie, Gratzer, 1966; Sarkar et al., 1967). In this region the optical rotation of the protein is very sensitive to the presence of α-helical structures which are responsible for the optical activity peak at 198 mμ, the minimum at 232 – 233 mμ, and the characteristic shoulder in the region of 210 mμ. It has been found that if the curve of the optical rotatory dispersion of ribosomal RNA is subtracted from the curve of the optical rotatory dispersion of the ribosomes, then the differential curve obtained, corresponding to the ribosomal protein, indicates the presence of α-helices. The total content of the α-helical structure in the ribosomal protein is tentatively estimated to be about 30%. In experiments with yeast ribosomes, it was found that in any case the isolation of the ribosomal protein from the ribosome leads to a complete disappearance of the secondary structure, i.e., evidently to denaturation of all or most of the ribosomal protein molecules (McPhie, Gratzer, 1966). However, a study of the optical activity of the ribosomal protein of *E. coli* indicated the possibility of a substantial preservation of α-helical structures in it after isolation from the ribosome (Sarkar et al., 1967).

It may be assumed that at least a certain portion of the molecules of ribosomal protein within the ribosome are globular-type molecules, containing α-helical regions.

5. Intraribosomal Packing

In addition to the heterogeneity of the ribosomal protein with respect to molecular weight and with respect to amino acid composition and sequence, it can also be shown that it is not homogeneous with respect to the mode and strength of its retention within the ribosome. Thus, the ribosomal proteins of *E. coli*, both of the 50 S subparticle and of the 30 S subparticle, can be separated at least into three groups, differently bound within the ribosome: 1. proteins that dissociate readily from the ribosomal particle under conditions of high salt concentration (for example, 3 to 6 M CsCl) in the presence of Mg^{++} ions; 2. proteins that dissociate at a high salt concentration, but far more slowly and only under the condition of a relatively low Mg^{++} content in the medium and removal of the dissociating protein from the reaction mixture; 3. proteins that do not dissociate under these conditions and can be removed from RNA only in the presence of a high concentration of salt (3 to 6 M) and urea (3 to 6 M) (SPIRIN et al., 1965; LERMAN et al., 1966; GARBER and BARULINA, 1968). Proteins that dissociate at high salt concentrations (in addition to CsCl, also LiCl, KCl, NH_4Cl, and other salts can be used in corresponding concentrations — 3 to 6 M) are denoted by us as *additional structural protein* of the ribosomes; it includes the first and second of the groups enumerated above (A-I and A-II) and comprises a total of about half the molecules of ribosomal protein. The nondissociating protein (the third of the enumerated groups) can be denoted as *basal structural protein* of the ribosomes (B); it comprises the other half of the molecules of ribosomal protein. Successive removal of the three enumerated groups of ribosomal protein (A-I, A-II, and B) leads to complete "stripping" of the ribosomal RNA.

Each group of proteins contains polypeptide chains of different basicity and exhibits many electrophoretic bands. Each of the groups of proteins both of the 50 S and of the 30 S subparticles exhibits its own characteristic band distribution profile, i.e., actually has its own characteristic set of polypeptide chains. It has been shown that the group of relatively readily dissociating proteins (A-I), comprising about 20% of the total ribosomal protein of *E. coli* (MESELSON et al., 1964; SPIRIN et al., 1965; LERMAN et al., 1966) includes the acidic proteins of the ribosome, but also contains basic proteins, which are characteristic only of this group (TRAUB et al., 1966; NOMURA, TRAUB, 1966). On the whole, the proteins that dissociate in high salt concentrations (additional structural protein, A-I+A-II) (SPIRIN et al., 1965; LERMAN et al., 1966; MARCOT-QUEIROZ, MONIER, 1966) include all the acidic proteins of the ribosomes, part of the weakly basic proteins, part of the moderately basic proteins, and probably all of the strongly basic proteins (MARCOT-QUEIROZ, MONIER, 1966; ATSMON et al., 1967; GARBER and BARULINA, 1968). Correspondingly, the "nondissociating" group of proteins (basal structural proteins, B) consists mainly of the weakly basic and moderately basic polypeptide chains.

The different strength of retention of the groups of ribosomal protein within the ribosome, as can be seen, is *by no means determined unambiguously by the basicity* of the polypeptide chains (GARBER and BARULINA, 1968). Moreover, it is found that during dissociation, the proteins behave not as independent molecules, but as *groups of molecules*, i.e., they dissociate more or less *cooperatively* (SPIRIN et al., 1965; LERMAN et al., 1966). All this is an indication of the complex nature of the intraribosomal packing of the proteins. Of course, the matter is hardly limited to electrostatic attraction between the amino groups of the proteins and the phosphates of RNA or any other simple electrostatic interactions. A large, if not decisive contribution to the retention of proteins within the ribosome is probably made by the *protein-to-protein interactions and their mutual organization*. Unfortunately, however, nothing is yet known about the concrete nature of this organization within the ribosome.

6. Summary

The ribosome contains several tens of molecules (polypeptide chains) of structural ribosomal protein. The bacterial 70 S ribosome contains about 60 polypeptide chains with an average molecular weight of about 20,000. The animal 80 S ribosome evidently possesses up to 150 polypeptide chains.

The polypeptide chains of the ribosomal protein are heterogeneous with respect to molecular weight; among them are both chains of about 100 amino acid residues and relatively long chains of up to 300 to 400 amino acid residues; a substantial portion of the protein is represented by chains of 100 to 150 amino acid residues.

Most of the polypeptide chains of the ribosomal protein are of a basic character due to a certain predominance of basic amino acids over acidic amino acids. On the whole, the ribosomal protein is highly heterogeneous with respect to amino acid composition and sequence, representing a complex assortment of various polypeptide chains.

Preliminary data indicate that some of the protein molecules within the ribosome can be characterized at least partially by a globular conformation and by the presence of α-helical regions.

The mutual packing of the molecules of ribosomal protein within the ribosome is evidently complex and is characterized by the presence of several cooperative groups of interdependent molecules. The retention of these groups within the ribosome differs in strength and evidently in nature as well. The retention of structural protein within the ribosome is scarcely limited to electrostatic interactions of it with RNA; a large, if not decisive contribution is probably made by protein to protein interactions and the mutual organization of the protein molecules (polypeptide chains).

IV. Structural Transformations of Ribosomes

At the present time, three types of structural transformations of ribosomal particles are known: 1. reversible dissociation of the ribosome into two subparticles; 2. unfolding of subparticles; 3. disassembly ("stripping") of subparticles.

1. Reversible Dissociation-Association

The phenomenon of dissociation of the ribosomes into two unequal subparticles was shown and understood in the very first works on the isolation and characterization of ribosomes from plant, bacterial, and animal objects (Chao, 1957; Ts'o et al., 1958; Tissières, Watson, 1958; Tissières et al. 1959; Ts'o, Vinograd, 1961). This phenomenon expresses itself in the fact that when the Mg^{++} ion concentration in the medium is lowered (or when the concentration of competing monovalent cations is increased), the initial 70 S or 80 S ribosome dissociates into its two component subparticles:

$$70\ S \rightleftharpoons 50\ S + 30\ S;$$

$$80\ S \rightleftharpoons 60\ S + 40\ S.$$

This dissociation is reversible, i.e., the addition of Mg^{++} ions to the dissociated subparticles promotes their specific association, in a 1:1 ratio, into the 70 S or 80 S particle, respectively. In the association of the subparticles into a complete biologically active ribosome (70 S or 80 S), the Mg^{++} ions can be replaced by Ca^{++} ions (but not by Be^{++}, Sr^{++}, or Ba^{++}) in an equivalent concentration (CHAO, 1957; Ts'o, 1958; GORDON, LIPMANN, 1967). Evidently the maintenance of a definite concentration of Mg^{++} and/or Ca^{++} in the medium and in the ribosome is vital for the connection between the larger and smaller subparticles. The connection between the subparticles is also stabilized by polyamines, especially spermidine (COHEN, LICHTENSTEIN, 1960; PESTKA, 1966).

Dissociation of the Bacterial 70 S Ribosomes. The bacterial ribosomes, in particular the ribosomes isolated from cells of *E. coli* and not subjected to special purification, exist primarily in the associated (70 S) state in the presence of 0.005 M to 0.01 M Mg^{++} in 0.01 M tris-buffer, pH 7.4; a 20-fold lowering of the Mg^{++} concentration — to 0.00025 M or 0.0005 M — causes them to dissociate into the 50 S and 30 S subparticles; when tris-buffer is replaced by phosphate buffer (0.01 M), the dissociation goes to completion even at higher concentrations of Mg^{++} — 0.001 M (TISSIÈRES et al., 1959). Thus, the associated (70 S) state of the ribosomes isolated from cells of *E. coli* is more or less stable at concentrations of $Mg^{++} \geq 0.005$ M (if the concentration of monovalent cations is of the order of 10^{-2} M). The dissociated state (50 S + 30 S) is stable at 10^{-4} M to 10^{-3} M Mg^{++}. At intermediate Mg^{++} concentrations, both states coexist (70 S + 50 S + 30 S). The presence of monovalent cations in a concentration of about 0.1 M somewhat shifts the reaction in the direction of dissociation (WATSON, 1964).

In addition to the specific association of the 50 S and 30 S subparticles into the 70 S ribosome, high concentrations of Mg^{++} (especially 0.01 M and higher) can promote dimerization of the 70 S ribosomes into 100 S particles (TISSIÈRES et al., 1959), as well dimerization of the 50 S subparticles (HUXLEY, ZUBAY, 1960) and dimerization of the 30 S subparticles (DAHLBERG, HASELKORN, 1967). At higher concentrations of Mg^{++}, there may be a further aggregation of the particles. The presence of K^{+} ions in a concentration of 0.05 M or 0.1 M inhibits these nonspecific effects of dimerization and aggregation (WATSON, 1964; DAHLBERG, HASELKORN, 1967).

Analogous dimerization of the 80 S ribosomes into 110 S or 120 S particles and of the 40 S subparticles into 60 S subparticles may also be observed in the case of the ribosomes of higher organisms (Ts'o et al., 1958; HAMILTON, PETERMANN, 1959; PETERMANN, 1960; Ts'o, VINOGRAD, 1961; LEDERBERG, MITCHISON, 1962; TASHIRO, SIEKEVITZ, 1965).

Dissociation of the 80 S Ribosomes. The 80 S ribosomes of higher organisms, and especially animal ribosomes, may differ greatly from the bacterial 70 S ribosomes with respect to the conditions of dissociation into the subparticles. The connection between the subparticles in the 80 S ribosomes may be far more stable. Usually, to induce dissociation of the 80 S ribosomes into subparticles, it is not sufficient to simply lower the concentration of Mg^{++} ions in the medium by one or two orders of magnitude, as in the case of the 70 S ribosomes, but most often the ribosomes must be placed in a medium without any Mg^{++} ions at all. But even in this case, dissociation usually does not occur; a second condition should be either a relatively high concentration of monovalent cations (for example, K^{+}), or an increased pH of the solution (7.5 to 8.5), or the presence of anions of the type of phosphate or carbonate, possessing an affinity for divalent cations. If strong chelating agents for the removal of Mg^{++} (of the type of EDTA or pyrophosphate), or extreme pH values (9 to 9.5), or high concentrations of phosphate or carbonate (0.1 M), or very high concentrations of monovalent cations (for example, 0.5 to 0.7 M KCl) are not used, then the 80 S ribosome can be dissociated under relatively mild conditions into the larger and smaller subparticles with sedimentation coefficients of about 60 S and 40 S, respectively. Thus, CHAO (1957) has described the complete dissociation of yeast ribosomes according to the type 80 S $\rightarrow$ 60 S + 40 S in 0.01 M phosphate buffer, pH 7.5. It was found to be possible to dissociate the ribosomes of pea seedlings according to the scheme 79 S $\rightarrow$ 58 ($\pm$ 2) S + 38 ($\pm$ 2) S by increasing the pH from 6.5 to 7.5 in phosphate buffer with an ionic strength of 0.05, or by increasing the ionic strength of the phosphate buffer pH 6.5 above 0.05, or under the action of 0.7 M KCl in the absence of phosphate, or, finally, simply by prolonged dialysis against phosphate buffer without Mg^{++} (Ts'o et al., 1958). Animal ribosomes, as is shown by the example of ribosomes of Novikoff rat hepatoma, not pre-treated with deoxycholate, also dissociate according to the scheme 81 S $\rightarrow$ 64 S + 45 S, for example, in 0.005 M phosphate buffer containing 0.0005 M $MgCl_2$ and 0.05 M NaCl (KUFF, ZEIGEL, 1960). In exactly the same way, ribosomes from rabbit reticulocytes dissociate according to the scheme 78 S $\rightarrow$ 58 S + 40 S during dialysis against a buffer with KCl containing no Mg^{++} (TS'O, VINOGRAD, 1961). The exact values of the sedimentation coefficients of the subparticles obtained, cited by different authors for different ribosomes, vary somewhat — from 58 to 64 S for the larger subparticle and from 38 to 46 S for the smaller subparticle.

In many cases, pre-treatment of the preparations with deoxycholate substantially changes the stability of the ribosomes with respect to dissociating influences, and possibly the nature of the ribosomes themselves, as a result of which the visible picture of dissociation is substantially modified (see, for example, KUFF, ZEIGEL, 1960). There is a suspicion that deoxycholate may split out some most labilely bound protein portion from the ribosome. Therefore, a special consideration will be required for those cases when deoxycholate (or some other detergent) is used to obtain the 80 S ribosomes. In exactly the same way, the use of more drastic conditions for the removal of Mg^{++} ions from the ribosome, including the use of chelating agents of the type of EDTA and pyrophosphate, causes structural changes in the ribosomes which go beyond the limits of simple dissociation into subparticles; dissociation in this case is evidently accompanied by partial unfolding (loosening) of the subparticles, and possibly even by the splitting off of part of the protein. Here we shall merely indicate that in the dissociation of the 80 S ribosomes with chelating agents, the sedimentation coefficients of the subparticles formed are substantially lower than those just cited, and for all the cases studied, independent of the object, they are 43 to 50 S for the larger subparticle and 24 to 32 S for the smaller subparticle (Ts'o et al., 1958; HAMILTON, PETERMANN, 1959; KUFF, ZEIGEL, 1960; LAMFROM, GLOWACKI, 1962; TASHIRO, SIEKEVITZ, 1965). This means that in such cases the dissociation of the 80 S ribosome into subparticles is most likely accompanied by other types of structural transformations — partial unfolding and perhaps slight disassembly — and does not represent a phenomenon in the pure form.

Influence of Dissociation on the Conformation of Ribosomal RNA and Protein. Since the associated state of the ribosome evidently exists on account of some sort of interaction of the surfaces of the two subparticles being in contact with one another, naturally the question arises of whether dissociation is accompanied by any conformational rearrangements within the subparticles themselves. These conformational rearrangements might be either a consequence of dissociation or, on the contrary, might be induced primarily by a decrease in the concentration of Mg^{++} and might thus be the direct cause of dissociation. A final answer to this question cannot yet be given. However, a study of the optical rotatory dispersion of the ribosomes and their subparticles gives evidence that whatever the case may be, there are no great visible changes in the secondary structure of the ribosomal RNA or ribosomal protein as a result of dissociation (McPhie, Gratzer, 1966; Sarkar et al., 1967). It may be supposed that the subparticles specifically interact with one another with their definite fixed regions on their flattened surfaces, without undergoing any reorganizations during the process of dissociation and association.

Subparticle Regions Responsible for Association. From electron microscopic observations it is evident that each subparticle interacts with another only with *a strictly definite region* of its surface (henceforth this region of the surface of the subparticle will be denoted as the *contacting surface*). There is one contacting surface on each subparticle. The contacting surface of the larger subparticle possesses a specific affinity only for the contacting surface of the smaller subparticle, and therefore, from a mixture of 50 S and 30 S (or 60 S and 40 S) subparticles, if we exclude nonspecific dimerization and aggregation, only the 50 S — 30 S (or 60 S — 40 S) associates always arise. Consequently, the contacting surfaces of the two subparticles bear specific sites or groups, possessing selective affinity for one another. In principle, such specific "complementariness" can be ensured either by complementary (Watson-Crick) interaction of the single-stranded sections of ribosomal RNA, situated on the contacting surfaces and specially performing this role, or protein-to-protein interactions. Any non-specific interactions of the purely ionic, metal-complexing, or other types, evidently are rather improbable here, although they may of course play a secondary role. In the light of this, the associating role of Mg^{++} or Ca^{++} may be reduced chiefly to screening of the negatively charged phosphate and carboxyl groups on the contacting surfaces, and thereby may consist of permitting close contact for specific interaction.

Which of the two most probable types of interaction — complementary pairing of sections of the polynucleotide chains of RNA or specific protein-to-protein contacts — is actually realized and plays the main role in the association of the subparticles cannot yet be stated. The following data support the first possibility, the complementary interaction of RNA: 1. isolated 23 S and 16 S ribosomal RNA's of *E. coli* can specifically form a complex with one another in the presence of 0.01 M Mg^{++} [Marcot-Queiroz, Monier, 1965 (1, 2)]; 2. treatment of the 50 S and 30 S subparticles with formaldehyde, nitrous acid, or perphthalic acid, i.e., with agents that act on the amino groups of RNA, deprives them of their ability to reassociate into the 70 S particles; at the same time, treatment with dinitrofluorobenzene under conditions in which it blocks the amino groups of the protein but does not react with RNA does not deprive the subparticles of their ability to reassociate (Moore, 1966). However, there are also some indications in support of protein-to-protein interactions: 1. very mild treatment of the ribosomes with proteolytic enzymes causes them to dissociate into subparticles; the subparticles, slightly treated with proteases, are incapable of associating into 70 S (or 80 S) ribosomes (MORGAN et al., 1963; ZAK et al., 1966); 2. treatment of *E. coli* ribosomes with reagents that act upon the sulfhydryl groups of the protein causes them to dissociate into subparticles (TAMAOKI, MIYAZAWA, 1967); 3. removal of part of the ribosomal protein with the aid of high concentrations of CsCl leads to the formation of protein-deficient derivatives that are incapable of associating in the presence of 0.01 M Mg^{++} (LERMAN et al., 1966). It is possible that only both types of interactions *cooperatively* ensure proper retention of the two subparticles together, and neither of them is sufficient alone.

Stability of Association. Everything that has been stated above concerning the conditions of dissociation and reassociation pertains to ribosomes isolated from cells in the form of the 70 S or 80 S particles. These complete ribosomal particles are present in the cells as functioning ribosomes, and as a rule, in the form of polyribosome structures. Isolation of ribosomes from cells is evidently accompanied by substantial fragmentation of the messenger RNA chains uniting them. However, even after isolation and purification, the 70 S and 80 S ribosomes can retain short mRNA fragments, aminoacyl-tRNA and peptidyl-tRNA molecules, and also possibly certain protein factors, participating in the organization of the active associates. The connection between the two subparticles in such ribosomes is found to be rather stable in the presence of 0.01 M Mg^{++}. The conditions of dissociation were described above.

However, it has been shown on such ribosomes from *E. coli*, that if they dissociate into subparticles, and then reassociate again into complete 70 S ribosomes, the connection between the subparticles in the reassociated ribosomes is found to be weaker under identical conditions of salt environment and Mg^{++} content than in the nondissociated ribosomes (PESTKA, 1966). The impression is created that when the original ribosome dissociates, some sort of boundary component between the subparticles, which plays an important role in stabilizing the association between the subparticles, is lost.

If, after dissociation of the complete 70 S ribosome, the subparticles formed are subjected to thorough purification, for example, by washing with 0.5 M NH_4Cl, then the purified subparticles will associate with one another quite weakly; even in the presence of 0.01 M Mg^{++} in the reaction mixture only a fraction of 70 S particles will be observed, i.e., the equilibrium of the reaction will be shifted more in the direction of dissociation (GAVRILOVA, SPIRIN, unpublished; see also NAKADA, KAJI, 1967).

There are independent data indicating that free (nonassociated) 50 S and 30 S subparticles, which *in vitro* are incapable of forming any stable 70 S ribosomes even at high (0.01 M) concentrations of Mg^{++}, can be isolated from the living bacterial cell (GREEN, HALL, 1961; MANGIAROTTI, SCHLESSINGER, 1966; NAKADA, KAJI, 1967). The indicated "native" subparticles are indistinguishable from the subparticles contained in the 70 S ribosomes, either according to RNA or according to the principal set of protein components (NAKADA, KAJI, 1967). In the complete protein-synthesizing system, a mixture of these free subparticles gives biologically active ribosomes. Evidently they associate into 70 S ribosomes through the formation of an active complex with mRNA and aminoacyl-tRNA, i.e., *only association with the components of the protein-synthesizing system induces a shift of the equilibrium of the reaction in the direction of more stable association of the two ribosomal subparticles with one another.*

It may be that *in vivo* only the *functioning* ribosomes (chiefly the ribosomes contained in the polyribosomes) are present in the form of *stable* 70 S (or 80 S) particles.

Recently the data were reported that one of the main components stabilizing the association of subparticles is a peptidyl-tRNA bound to a 50 S subparticle (SCHLESSINGER et al., 1967). The pool of *nonfunctioning* ribosomal particles may be represented as an equilibrium mixture

$$50\ S + 30\ S \rightleftharpoons 70\ S,$$

where the association between the subparticles is extremely *unstable*, and therefore the equilibrium is substantially shifted in the direction of the free subparticles (see MANGIAROTTI, SCHLESSINGER, 1966). Association with the components of the protein-synthesizing system stabilizes the connection between the subparticles, shifting the equilibrium in the direction of association. Initiation of translation and appearance of peptidyl-tRNA in the ribosome makes the association of the subparticles practically irreversible under physiological conditions. Termination of translation, i. e., release of the polypeptide from the ribosome may again lead to a labilization of the connection between the subparticles and their return to the equilibrium pool of nonfunctioning particles. In any case, in a cell-free protein-synthesizing system, even in the presence of a high Mg^{++} concentration (0.016 M), the 70 S ribosomes can dissociate readily and exchange their subparticles (TAKEDA, LIPMANN, 1966).

In all probability, the lability of the association of the subparticles in the ribosome has some great biological significance, and there must be not simply an affinity of the two subparticles for one another, but an affinity of a *quite definite strength*, neither too small *nor too great*, for the ribosome to carry out its functons. The ribosomes of halophilic bacteria, in particular, the extremely halophilic organism *Halobacterium cutirubrum*, can serve as a clear illustration of this (BAYLEY, KUSHNER, 1964; BAYLEY, 1966). This organism lives in concentrated salt solutions, as a result of which its intracellular salt concentration is very great; probably a concentration of no less than 3 to 4 M K^+ and other monovalent cations and a 0.1 M concentration of Mg^{++} is maintained in it. Correspondingly, its 70 S ribosomes are stable in the presence of 1 to 4 M KCl with 0.1 to 0.4 M $MgCl_2$. In the case of normal ribosomes, such concentrations of Mg^{++} ions, even in the presence of high ionic strength, should lead to firm fastening of the subparticles and even to aggregation of the ribosomes. However, in this organism the ribosomal proteins have been found to be far more acidic than in the usual bacteria, which evidently ensures, in particular, lability of the association of the ribosomal subparticles even under these strongly "stabilizing" conditions. In the presence of 4 M KCl and 0.1 M $MgCl_2$, the 70 S ribosomes of halophilic bacteria lie on the border of stability, and lowering the concentration of the indicated salts below this level leads to dissociation into the 50 S and 30 S subparticles. Thus, if the ribosomes have to work at a high concentration of salts and Mg^{++}, Nature still ensures lability of the association of their subparticles — chiefly through the corresponding changes in the ribosomal protein.

On the whole, the construction of a biologically active ribosome from two unequal subparticles, and most important, the relative *lability of their association* is a universal feature of all ribosomes without exception. This must have some fundamental importance for the functioning of the ribosome, which we do not yet know. We can only surmise that this ensures mobility of the two ribosomal subparticles relative to one another in the process of translation, and that the periodic relative movements of the subparticles are an important factor in the functioning of the ribosome.

2. Unfolding

To ensure a native compact state of each ribosomal subparticle in solution, the effective electrostatic repulsion of the anionic groups (phosphates and carboxyls) within the particle must not exceed some definite value. Usually this is achieved by shielding of the intraribosomal negative charges by the numerous Mg^{++} ions bound in the ribosomes and, evidently, to a definite degree by the ions of polyamines as well.

High concentrations of monovalent cations in the medium may lead to a competitive displacement of part of the Mg^{++} ions (and polyamines) from the ribosomes. At definite concentrations of salts of monovalent metals this partial displacement of the intraribosomal cations is compensated for by the shielding effect of the high concentration of free monovalent cations, and the ribosomal particles do not lose their original compactness. However, if after such treatment of the ribosomes, they are transferred to conditions of lower ionic strength without Mg^{++} ions in the medium, these Mg^{++}-depleted ribosomal particles exhibit a sharp, abrupt transformation to a less compact form [Spirin et al., 1963; Gavrilova et al., 1966 (1)]. This transition is denoted as *unfolding*.

Such unfolding of the ribosomal particles can also be produced by direct removal of Mg^{++} ions from the bacterial ribosomes — for example, with the aid of chelating agents (EDTA) (Gesteland, 1966).

On the basis of an analysis of the reported data, it may be concluded that certain stages of the process of unfolding, accompanied by a decrease in the sedimentation coefficients of the ribosomal subparticles, have been periodically observed on bacterial ribosomes by various authors, although these phenomena have been treated differently (Weller, Horowitz, 1964; Rodgers, 1964; Cammack, Wade, 1965). It may be that Elson (Elson, 1961, 1964) also dealt with a loosening of the 50 S particle (50 S → 40 S), although in his experiments there might have been elements of disassembly (see below). The conversion of the yeast 60 S subparticle to the 50 S component during dialysis against water (Morgan, 1962) also most likely represented a loosening, although here also the presence of accompanying partial dissasembly cannot be categorically denied.

It is likely that in the case of drastic dissociation of animal 80 S ribosomes in the presence of pyrophosphate or EDTA, dissociation into the subparticles may be accompanied by a certain unfolding of the subparticles, so that their sedimentation coefficients drop from 60 S and 40 S to 30 S—50 S and 20 S—30 S, respectively (Tashiro, Siekevitz, 1965; Gould et al., 1966; Petermann, Pavlovec, 1966). The loosening of the 60 S and 40 S subparticles in magnesium-free medium under low ionic strengths was probably observed even earlier (Ts'o, Vinograd, 1961).

The unfolding of the ribosomal subparticles, discovered and studied on the ribosomes of *E. coli* [Spirin et al., 1963; Gavrilova et al., 1966 (1, 2); Gesteland, 1966] represents a multi-stage process. Thus, at first the original compact 50 S subparticle *abruptly* transforms into the 35 S form characterized by the same molecular weight and RNA-to-protein ratio, but by a less compact state than the 50 S form. Then, when Mg^{++} is further removed, or the ionic strength is lowered, the 35 S form changes into a still less compact form, the 22 S — also abruptly, without intervening states, according to the "all-or-none" principle. Despite the conservation of the complete content of ribosomal protein, the 22 S form behaves like a typical polyelectrolyte of the type of free RNA in solution and shows a more or less gradual decrease in its sedimentation coefficient, i.e., gradually unfolds, when the ionic strength is lowered in the absence of Mg^{++} ions.

The unfolding of the 30 S subparticle proceeds analogously, through definite discrete stages.

Schematically, the process of unfolding can be represented as follows:

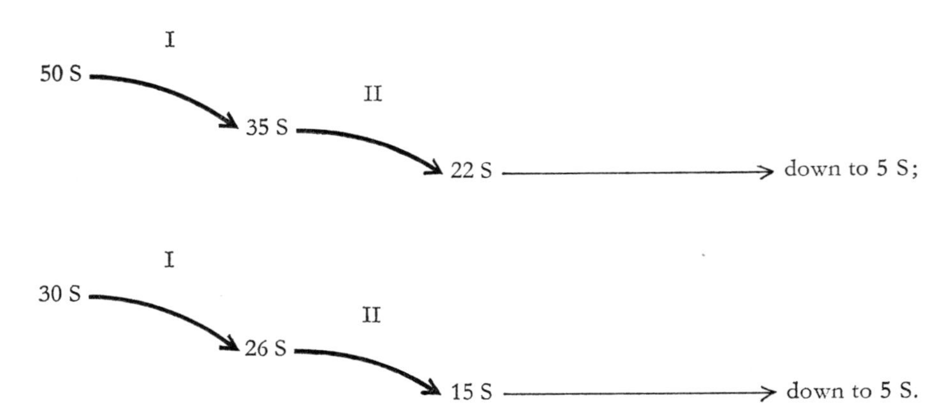

Each abrupt stage of the unfolding — I and II — proceeds within a rather *narrow interval* of variation of the ionic conditions. For example, if unfolding is accomplished by a reduction of the ionic strength in the medium, then stage I occurs chiefly within the interval of ionic strength (NH_4Cl or KCl) from 0.3 to 0.2, while stage II occurs chiefly within the interval from 0.1 to 0.01 [Gavrilova et al., 1966 (1)].

The unfolding of the subparticles, if it did not go too far, may evidently be reversible. Reversal — complete or partial — is achieved by the addition of Mg^{++} ions.

Hence, a characteristic feature of the unfolding process is that during the process three successive *discrete* states of the ribonucleoprotein particle can be discerned (for example, 50 S, 35 S, and 22 S). Transition from one state of the particle to another is an abrupt, cooperative transformation, proceeding according to the "all-or-none" principle, without intervening stages. Thus, if we take the liberty of speaking of three phase states of the ribonucleoprotein particle, then each stage of the transformation may be considered as a phase transition.

The final — loose — state of the ribonucleoprotein particle is characterized by the absence of any definite, fixed value of the sedimentation coefficient. It depends greatly upon the ionic strength of the solution. The particles in this state behave as typical random polyelectrolytes, gradually increasing their molecular volume with decreasing ionic strength. Therefore, this state of the ribonucleoprotein particle can logically be considered as a *random* state of the *ribonucleoprotein strand*, where the frame or skeleton is formed by a strand ("rod") of RNA (see above, Section II, 6), with protein molecules attached to it.

The original state of the ribosomal particle in this case may be considered as some sort of compact packing of the indicated ribonucleoprotein strand. The compactness of this packing is evidently ensured by the interaction between sections of the ribonucleoprotein strand. It is very probable that protein-to-protein interactions play the basic role here.

The cooperative nature of the transition of the original — compact — state of the ribosomal particle into the state of a loose random ribonucleoprotein strand may indicate the existence of a definite regularity in the intraribosomal compact packing of the strand sections. Then, as a result of the first stage of the transformation, some cooperative system of protein-to-protein interactions between sections of the ribonucleoprotein strand is disturbed, and the ribosomal particle is converted from a compact regular state to a more loosened intermediate state, which, however, retains part of its regularity. As a result of the second stage of the transformation, the remaining interactions among sections of the strand are disturbed, and the ribonucleoprotein particle passes into the state of the unfolded random ribonucleoprotein strand.

3. Disassembly

As has already been noted, high concentrations of monovalent cations exhibit a dual effect upon the intraribosomal electrostatic forces: a) weakening of all the electrostatic interactions, including both electrostatic repulsion (chiefly between the RNA phosphates) and electrostatic attraction (possibly between the RNA phosphates, on the one hand, and the basic groups of the proteins, on the other); b) competitive displacement by the monovalent cations of the bound Mg^{++} ions in the ribosome. Apparently it is the weakening of the electrostatic interaction between the protein and RNA and the displacement of Mg^{++} ions that leads to the effect of the ribosomal protein becoming less firmly retained within the ribosomal particle under the action of high concentrations of monovalent salts. Usually, if the Mg^{++} concentration in solution does not exceed 0.01 M, then at salt concentrations of 1—2 M and above, dissociation of the protein from the ribosomal particle can already begin. The addition of Mg^{++} ions to the medium counteracts the dissociation of the protein molecules from the ribosomal particle. This is an indication that the retention of protein within the ribosome is ensured not only by simple electrostatic interaction of its basic groups with the RNA phosphates, but to a substantial degree, if not chiefly, *on account of some sort of organization with the participation of* Mg^{++} *ions* (see below).

An indication of the possibility of separating a definite, comparatively small portion of protein from the ribosomal particles was first obtained on ribosomes of *E. coli*, treated with high concentrations of CsCl (Meselson et al., 1964). Subsequently it was shown that successive "stripping" of the ribosomal particles of *E. coli*, i.e., stepwise splitting off of the protein in portions, in several discrete stages, can be produced with high concentrations of CsCl [Spirin et al., 1965, 1966; Lerman et al., 1966; Gavrilova et al., 1966 (2)]. It is this multi-stage process that is defined as the *dissasembly* of the ribosomal particles. Analogous stepwise disassembly can also be accomplished through the use of high concentrations (2 to 6 M) of NH_4Cl, KCl, and LiCl (Ivanov, Spirin, unpublished). The disassembly procedure elaborated for bacterial ribosomes was found to be successfully effective also for disassembly of animal ribosomal particles (Lerman, 1968).

Schematically the process of disassembly of the ribosomal particles of *E. coli* can be represented in the following way:

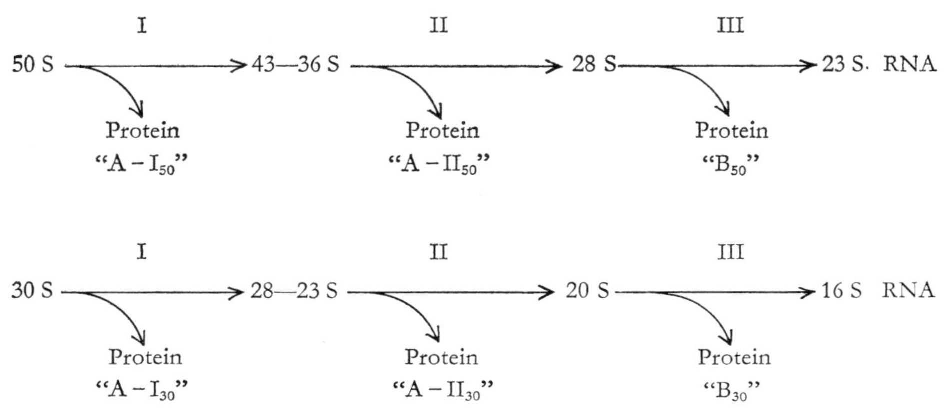

At the first stage of disassembly, a portion of the protein "A-I" ("additional structural protein I"), comprising about 20% of the total ribosomal protein, is stripped off. At the second stage, about 30% of the ribosomal protein — portion "A-II" ("additional structural protein II") dissociates. Finally, the third step, which occurs only in the case of complete removal of Mg^{++} ions with EDTA or in the presence of urea, means splitting off from RNA of the remaining 50% of the ribosomal protein, i.e., of portion "B" ("basal structural protein"). Correspondingly, the original ribosomal 50 S and 30 S subparticles contain 37% protein; the intermediate 43 S and 28 S particles (A-particles, or "intersomes") contain 30% protein (80% of the original protein content), while the final ribonucleoprotein 28 S and 20 S particles ("CM-like particles" or "minisomes") contain only about 20% protein (50% of the original protein content) (see Table 7).

Properties of the Intermediate 43 S — 36 S and 28 S — 23 S Particles (Intersomes). Protein-deficient derivatives of the 50 S and 30 S ribosomal subparticles can be produced by prolonged (equilibrium, about 36 h) CsCl density gradient *centrifugation* of the ribosomes of *E. coli* in the presence of a relatively high concentration of Mg^{++}, 0.02 M — 0.04 M (MESELSON et al., 1964; HOSOKAWA et al., 1966; STAEHELIN, MESELSON, 1966). The derivatives thus obtained were characterized by sedimentation coefficients of 40 to 42 S and 23 S in 0.01 M phosphate buffer with 0.001 M Mg^{++} or in 0.01 M tris-buffer with 10^4 M — 5×10^{-4} M Mg^{++}. According to indirect data, based on a calculation of the amount of radioactive sulfur in preliminarily labeled total protein and the protein split out, they contain 60 to 80% of the original protein content. The buoyant density in CsCl in the presence of 0.02 to 0.04 M Mg^{++} was 1.65 to 1.66 g/cm³, while the original ribosomal particles gave a buoyant density of about 1.61 to 1.62 g/cm³ under the same conditions.

Analogous particles can be obtained by prolonged (15 to 72 h) *incubation* of *E. coli* ribosomes in 5 M CsCl, but at a relatively low Mg^{++} concentration — about 0.002 M (LERMAN et al., 1966). In 0.01 M phosphate buffer with 0.001 M Mg^{++}, the particles obtained were characterized by sedimentation coefficients $S^0_{20,w}$ of either 42 — 43 S and 28 S, or, as a result of longer incubation in CsCl, 36 S and 23 S. Direct determinations of the amount of protein indicated that they contained about 30% protein, i.e., about 80% of the original protein content. The buoyant density in CsCl in the presence of 0.003 M Mg^{++} was 1.68 to 1.69 g/cm³; the original ribosomes possessed a buoyant density of 1.62 to 1.64 g/cm³ under the same conditions.

Table 7. *Terminology and properties of E. coli ribosomal particles and their natural precursors or artificial protein-deficient derivatives*

Properties	Ribosomes		Neosomes		Eosomes	
	Complete ribosome	Ribosomal subparticles	Inter-somes	Mini-somes	?	RNA
Sedimentation coefficient, Svedberg units	70	50 30	43—36 28—23	28($\pm$2) 20($\pm$2)	? ?	23 16
Protein content, %	37	37	$\approx$30	$\approx$20	Very little	0
Buoyant density in CsCl, g/cm³	1.61—1.64	1.61—1.64	1.65—1.69	1.72—1.74	?	$\geq$1.9

The indicated particles are denoted differently by different authors: A-component or "A-particles", "core particles", "subribosomal particles", "stripped particles", etc. We propose that they be denoted as *intersomes.*

Perhaps the intersomes can be subdivided into at least two subgroups, possibly differing somewhat in protein content: the subgroup of the 43 S and 28 S particles and the subgroup of the 36 S and 23 S particles.

The sedimentation coefficients of the intersomes evidently depend little upon the Mg^{++} ion concentration in solution, in any case, within the range from 10^{-4} to 10^{-3} M.

The intersomes, in contrast to the original ribosomal particles, are functionally inactive: they are incapable of binding mRNA and tRNA (Raskas, Staehelin, 1967; Nomura, Traub, 1966), nor, on the whole, can they carry out polypeptide synthesis *in vitro* (Hosokawa et al., 1966; Staehelin, Meselson, 1966; Nomura, Traub, 1966).

Properties of the "Minimal" Ribonucleoprotein 28 S and 20 S Particles (Minisomes). Derivatives even more deficient in protein can be obtained from the ribosomal particles or from the intersomes by centrifugation through concentrated (2—3 M) solutions of CsCl or LiCl with a low Mg content (2×10^{-3} M to 5×10^{-5} M) (Spirin et al., 1965, 1966; Lerman et al., 1966; Marcot-Queiroz, Monier, 1966). The particles obtained contain only about 50% of the initial protein content of the ribosomes. Correspondingly, their buoyant density in CsCl in the presence of 2×10^{-3} M Mg^{++} is 1.72 to 1.74 g/cm³. Their sedimentation coefficients are highly dependent upon the concentration of the particles and upon the concentration of Mg^{++} ions. The values of $S^0_{20,w}$ in the presence of 10^{-3} M Mg^{++} are about 28($\pm$2) S and 20($\pm$2) S; in the presence of 10^{-4} M Mg^{++} they are essentially lower, being about 25 S and 18 S. Increasing the ionic by the addition of monovalent ions also leads to an increase in the sedimentation coefficients. Evidently this reflects reversible conformational changes of the particles, depending upon the Mg^{++} ions and ionic strength. It was shown that the changes consist in the unfolding (loosening of the particles) upon lowering Mg^{++} concentration or ionic strength (Ivanov, Spirin, unpublished).

With respect to all the properties studied, the particles obtained are very similar or even practically indistinguishable from the natural ribosomal precursors. Particles of the ribosomal precursor type can accumulate in especially great amounts when protein synthesis is inhibited in bacterial cells with chloramphenicol ("CM-particles"), puromycin ("PM-particles"), or as a result of starvation of a required amino acid articles"), etc. That is why the artificially stripped derivatives of ribosomes consideration are sometimes denoted as "CM-like" particles.

Since the ribosomal protein cannot be further split off from these particles by the most varied methods, other than subjecting the ribosomal RNA to complete deproteinization, we propose that the indicated "minimal" ribonucleoproteins be called *minisomes*. It is proposed that the corresponding natural precursors of the ribosomes (25 – 30 S and 20 S neosomes and probably CM-particles, PM-particles, RC-particles, etc.) also be attributed to minisomes.

Characteristics of the Process of Disassembly. The most important and characteristic feature of the process of disassembly is that the splitting off of proteins from the ribosomal particle occurs not as a successive dissociation of numerous independent protein molecules, but by discrete groups, cooperatively. If, for example, a population of 50 S particles is taken into consideration then in the case of incubation in 3 to 6 M CsCl with 0.002 M $MgCl_2$, they are converted to 43 S particles, which are protein-deficient. Complete conversion requires no less than 40 h. But this lengthy process of conversion of a *population* of 50 S particles to a population of 43 S particles occurs in such a way that at each given moment each particle may either remain in the form of a 50 S particle or may be abruptly converted to a 43 S particle. In the latter case, seven to ten polypeptide chains of the structural protein are split off *at once* from the particle. There are no intervening stages. *For each particle* this means a process of the "all-or-none" type. Consequently, these seven to ten polypeptide chains (protein molecules) are somehow interdependent in a cooperative group within the ribosome (this does not necessarily mean the presence of a direct contact interaction among the given molecules).

Thus, it is possible to speak of the presence of several cooperatively organized groups of polypeptide chains (protein molecules) in each ribosomal subparticle. On the scheme, these groups will correspond, for example to portions A-I, or A-II.

It is also characteristic of the process of disassembly of the subparticles that both the 50 S and the 30 S subparticles behave *quite homologously*. Each stage of the splitting off of protein is accompanied by the loss of an equal proportion of the total protein content from the 50 S and from the 30 S subparticles. The loss of a given portion from the 50 S and 30 S subparticles occurs under the same conditions. The successive, protein-deficient derivatives are identical in buoyant density, i.e., in the RNA-to-protein ratio both in the 50 S series and in the 30 S series. This necessarily indicates a homology of structural organization of the two subparticles, the larger and the smaller.

Reversibility (Self-Assembly of Ribosomal Particles). The process of disassembly is found to be of a reversible character. In other words, under definite conditions the products of disassembly (at least, of incomplete disassembly) are capable of reassembling specifically into ribosomal particles possessing functional (protein-synthesizing) activity. Such a reconstitution of biologically active ribosomal particles occurs spontaneously and represents an example of *self-assembly* of a complex biological structure.

The first experimental demonstration of the possibility of *spontaneous self-assembly* of ribosomal particles *in vitro* were the experiments with the ribosomal precursors of the minisome type (Shakulov et al., 1963; Spirin, 1963). As has already been indicated an especially great amount of protein-deficient ribonucleoprotein precursors of the minisome type can be accumulated in the cell when protein synthesis is inhibited with antibiotics or as a result of starvation of a required amino acid. In particular, the minisome type particles denoted in this case as "chloromycetin particles" or CM-particles are accumulated under the action of chloramphenicol upon the bacterial cell (NOMURA, WATSON, 1959; KURLAND et al., 1962). It has been shown that CM-particles isolated from the *E. coli* cells inhibited with chloramphenicol, are capable *in vitro* of readily adding a protein, being converted to ribosome-like 50 S and 30 S particles (SHAKULOV et al., 1963; SPIRIN, 1963). Subsequently, it was shown that these experiments are successfully accomplished only when using ribosomal protein and that it is the "additional structural protein" of ribosomes that effectively associates with CM-particles building them up to ribosome-like 50 S and 30 S particles (LERMAN et al., 1967).

It is precisely these experiments that induced attempts to artificially obtain CM-like minisomes by "stripping off" the ribosomal particles and then to reverse the process and reconstitute the ribosomes (LERMAN et al., 1966; SPIRIN et al., 1966). Actually, the process of disassembly of ribosomal particles to minisome (CM-like) particles comprising 50% of the original protein content was found to be comparatively easily reversible. If this split protein ("additional structural protein", A-I + A-II) is added to the suspension of minisomes ("CM-like particles") obtained as a result of splitting off from ribosomal particles half of their structural protein, then in the buffer, in the presence of Mg^{++} ions, the 28 S and 20 S minisomes specifically reassemble with the protein, producing 50 S and 30 S ribosome-like particles (LERMAN et al., 1966) capable of functioning (synthesizing polypeptides) in the protein-synthesizing cell-free system (SPIRIN and BELITSINA, 1966; SPIRIN et al., 1966).

It was shown independently that the reconstitution of biologically active ribosomal particles can also be accomplished from the less stripped intermediate particles — from 43 S and 28 S intersomes seemingly comprised of 80% of the original protein content (HOSOKAWA et al., 1966; STAEHELIN and MESELSON, 1966).

Recently, attempts have been made to achieve a reversal of all the disassembly stages, i.e., to reconstitute ribosomes from the completely stripped RNA and corresponding fractions of the ribosomal protein. It appears that the incubation of 23 S ribosomal RNA with the "basal structural protein" (B) of 50 S subparticles under definite conditions leads to their specific assembly into 28 S ribonucleoprotein particles indistinguishable from the 28 S minisomes (SPIRIN et al., 1968). Hence, evidently the first and most difficult assembly stage has been carried out *in vitro* — the association of RNA with the first portion of polypeptide chains. The process of formation of such 28 S minisome-like particles strictly requires only the given set of polypeptide chains ("basal structural protein"); during the incubation of 23 S RNA with the "additional protein" of 50 S subparticles or with the "basal protein" of 30 S subparticles the 28 S homogeneous component is not formed. The reconstituted 28 S minisome-like particles can further incorporate the "additional protein" of 50 S subparticles and build themselves up to intersome-like or ribosome-like particles.

Therefore, it may be possible to achieve the reversal of all stages of the process of disassembly. Every stage of the reverse process (assembly) is of a spontaneous character. Consequently, as a whole, the process represents *in vitro self*-assembly. Every stage is accompanied by a specific association of a central matrix, RNA, minisome or intersome, with a concrete set of various polypeptide chains. Successive stages of the *in vitro* self-assembly of the ribosomal particles can be represented schematically in the following manner:

23 S RNA → 28 S RNP → 36 S — 43 S RNP → 50 S ribosome;

16 S RNA → 20 S RNP → 23 S — 28 S RNP → 30 S ribosome.

On the basis of some data it may be assumed that in the cell too, ribosomes are formed by self-assembly through the discrete stages of protein-deficient ribonucleoprotein precursor particles (neosomes) by means of their successive building up with the protein (BRITTEN et al., 1962). Actually, the ribosomal precursors (neosomes) analogous to the stripped particles, intersomes and minisomes examined above and obtained artificially from ribosomes, can be detected in the bacterial cell. According to the recent scheme of OSAWA et al. (OSAWA et al., 1967) the formation of ribosomes *in vivo* in *E. coli* passes through the following intervening stages:

23 S RNA → 30 S RNP → 38 S — 43 S RNP → 50 S ribosome;

16 S RNA → 21 S RNP → 30 S RNP → 30 S ribosome.

One cannot but notice a practically full coincidence of this independently formulated scheme of ribosome biogenesis *in vivo* with the schemes cited above, namely, the scheme of ribosome disassembly (p. 72) if we consider it the other way round, i.e., from right to left, and with the scheme of ribosome self-assembly *in vitro* (p. 76). It may now be assumed that ribosomes in the cell are assembled spontaneously from ribosomal RNA and numerous polypeptide chains of the ribosomal protein through the several discrete states according to the scheme demonstrating stepwise disassembly of the ribosomes, but with the opposite direction of this process.

The possibility of proper spontaneous assembly of numerous and various protein molecules (polypeptide chains) into a ribosomal particle means that there is a specific and exact recognition by every molecule of its place within the ribosome. In this connection, it is very probable that some specific organizing matrix directs assembly; the role of the matrix could be played by the ribosomal RNA chain. Thus, the following hypothesis of the *functional role of the ribosomal RNA* in the cell can be formulated: the chain of the ribosomal RNA serves mainly as some sort of *a matrix on which the molecules of the ribosomal protein can be specifically arranged;* hence, every RNA region with specific nucleotide sequence is recognized by a definite protein; thereby, the nucleotide sequence of ribosomal RNA determines the set and order of ribosomal proteins. Of course, the adding of the first portion of proteins during the process of assembly ("basal ribosomal protein") can be determined *directly* by the RNA nucleotide sequence, whereas, the adding of the subsequent portions of proteins ("additional ribosomal protein") should be determined most likely both by the RNA structure and the *protein molecules already attached,* i.e., in the latter case the ribonucleoprotein will serve as a matrix. The matrix role of ribosomal RNA may be apparent not only in the process of self-assembly but in the maintaining of the integrity and in the functioning of the assembled ribosomal particles as well. Certainly, the hypothesis of the matrix role of ribosomal RNA does not exclude the possibility of the direct functioning of some of its regions in binding the components of the protein-synthesizing system and in participation in some catalytic functions.

4. Conformational Stability of Ribosomes

A study of the processes of conformational transformations of ribosomes of the unfolding and disassembly type *in vitro* definitely indicates that each ribosomal particle represents a rather stable cooperative system. In order to induce transformations of unfolding or disassembly type, it is evidently necessary to go far beyond the limits of those influences that are possible under physiological conditions. The detected cooperativity in the organization of the ribosomal subparticles is an important feature which protects the ribosome from possible changes and distortions not only under varying conditions *in vivo*, but also during rather drastic isolation procedures. From the standpoint of the facts that are presently at our disposal concerning the cooperativity of the conformational transformations of the ribosomal subparticles, any visible changes in the morphology of the ribosomes in the cell in response to physiological influences seem rather improbable. That is also why, i.e., on the grounds of the observed cooperativity and stability of the conformation of the ribosomal particles, the doubts sometimes expressed, that in an isolated preparation we inevitably obtain ribosomes quite different from those present in the living cell, can to a certain degree be dispelled.

5. Some General Tentative Conclusions on the Quaternary Structure of the Ribosome

a) The quaternary structure of both ribosomal subparticles, the 50 S and 30 S is organized monotypically, homologously, according to the same principle, despite the difference in their size.

b) The quaternary structure of each ribosomal subparticle represents a rather compact packing of protein molecules and RNA regions.

c) Compactness is ensured by the presence of close intraribosomal interactions. These are not only interactions of RNA with protein, but perhaps in the first place, close protein-to-protein interactions.

d) An important role in the intraribosomal interactions ensuring compactness of the particles is played by Mg^{++} ions bound within the ribosome. This role of Mg^{++} consists, in the first place, of a neutralization of phosphates and carboxyls, which suppresses intraribosomal electrostatic repulsion, and thereby permits protein-to-protein interactions, perhaps including some of a hydrophobic nature; in the second place, that it can and probably does form direct Mg^{++}-bridges within the ribosome, creating a system of intraribosomal cross-links.

e) There is a definite cooperativity in the organization of the intraribosomal interactions. Even the interaction of protein with RNA in the ribosome cannot be considered as a simple electrostatic interaction of independent protein molecules with the phosphate groups of RNA, but only as a cooperative system of polypeptide chains (protein molecules) and RNA regions, interacting with one another.

f) The presence of cooperativity in the intraribosomal interactions may reflect (although not necessarily) the existence of a definite order in packing of intraribosomal elements, i.e., a definite regularity in the quaternary structure of the ribosomal particles.

Part Two

Functioning of the Ribosome

I. Components of the Protein-Synthesizing System

The ribosome is the apparatus of protein synthesis. To bring about protein synthesis, the apparatus should be equipped with: 1. a *template polynucleotide*, the role of which in the cells is played by natural messenger RNA (mRNA), while in cell-free systems their role may also be played by various synthetic polyribonucleotides; 2. amino-acylated adaptor or transfer RNA's *(aminoacyl-tRNA)*; 3. a set of "*transfer*" *factors*, as well as "*chain initiation factors*" and "*chain termination factors*" of a protein nature; 4. GTP; 5. appropriate *inorganic cations* — divalent, Mg^{++} or Ca^{++}, and monovalent, K^{+} or NH_4^{+} — in a definite concentration. All these represent the *components of the protein-synthesizing system*. The conditions set by the system for the state of each of the components are described below.

We should make an important reservation at the outset: in all the exposition to follow, pertaining to the components of the protein-synthesizing system and to the functioning of the ribosome itself, we will be speaking almost exclusively of bacterial systems and bacterial ribosomes, chiefly of the system and ribosomes of *Escherichia coli* as an example. Systems including the ribosomes of higher organisms have not yet been sufficiently studied, and the comparatively scanty data available do not contain any essentially new information in comparison with bacterial systems. We cannot yet say where the basic difference between bacterial and animal systems lies and how far their similarity extends. On the whole, however, the components enumerated above are evidently common to all systems.

1. The Ribosome

It is known that the complete functioning ribosome consists of two, loosely associated subparticles — the larger (50 S) and the smaller (30 S) (Fig. 13). Only the *complete* ribosome (70 S), consisting of *both* the subparticles, can bring about the process of translation in the protein-synthesizing system. Neither of the ribosomal subparticles individually, the 50 S or the 30 S, can replace the complete ribosome in the synthesis of the polypeptide chain of a protein (Gilbert, 1963; Pestka, Nirenberg, 1966).

It should be kept in mind, however, that the formal integrity of the ribosome, established by physical methods, cannot always serve as a criterion of functional activity of a ribosome preparation in a protein-synthesizing system. Thus, in 1962 the phenomenon of so-called "latent degradation" of the ribosomes was discovered

(Shakulov et al., 1962), in which the chain of ribosomal RNA within the ribosome was broken at several points, as a result of the action of endogenous nuclease factors or exogenous ribonuclease, but this did not affect the visible formal integrity of the particles (molecular weight, dimensions, shape). Such ribosomes with latent degradation of RNA are found to be inactive in protein synthesis. Formally integral ribosomes with damaged ribosomal RNA are incapable of interacting normally with other components of the protein-synthesizing system — in particular, with tRNA (Kaji, Kaji, 1965). In a study of the effects of irradiation on the ribosomes, it was shown that even a single break in the ribosomal RNA within the ribosome is sufficient to deprive it of biological activity in the protein-synthesizing system (Kućan, 1966).

In exactly the same way, the ribosome may not lose its formal integrity after very mild treatment with proteolytic enzymes (Zak et al., 1966). In this case the ribosome loses its ability to interact with the template polynucleotide and, consequently, despite its apparent integrity, becomes inactive in the protein-synthesizing system. Nonenzymatic removal of a portion of the structural proteins, for example, with the aid of high concentrations of CsCl deprives the ribosome of its functions essential for protein synthesis: i.e., of the ability of the subparticles to be retained in an associated (70 S) state, to bind with the template polynucleotide and with tRNA, and on the whole, to incorporate amino acids into the peptide chain (Lerman et al., 1966; Nomura, Traub, 1966; Raskas, Staehelin, 1967).

Hence, to serve as an apparatus for translation, i.e., for all round functioning, the ribosomal particles should evidently satisfy the following compulsory requirements: a) be able to associate into the form of 70 S particles, i.e., to form a specific couple consisting of two unequal, loosely associated subparticles, the 50 S and 30 S; b) contain an intact chain of ribosomal RNA; c) contain a complete set of intact structural ribosomal proteins. In other words, the ribosome should retain a complete native structure.

2. The Template Polynucleotide

In order to interact with the ribosome and perform its template role in the protein-synthesizing system, the template polynucleotide should evidently satisfy a single necessary condition: it should not possess a perfect secondary structure or a large fraction of especially stable cooperative helical regions. Thus, neither the usual native DNA's nor double-stranded viral RNA's, nor helical complexes of synthetic polyribonucleotides of the type of the polyA – polyU complex, as well as synthetic polynucleotides with an especially large G content, possess template activity, and they are generally incapable of combining with the ribosomes [Takanami, Okamoto, 1963 (1, 2); Nirenberg, Matthaei, 1961; Singer et al., 1963; Nirenberg et al., 1963].

At the same time it is well known that natural mRNA's and single-stranded viral RNA's which are ideal templates, possess a rather developed secondary structure (see Spirin, 1963; Bautz, 1963; Haselkorn et al., 1963). This secondary structure represents a set of relatively short and relatively unstable helical regions, formed due to pairing of adjacent sections of the chain with one another (Fig. 14). The total helical content is high (up to 70 – 80% of all the nucleotides). Nonetheless, this does not prevent them from serving as a template in protein synthesis. Evidently, in order to associate with the ribosome, the polynucleotide primarily must possess single-stranded, nonhelical regions. It is especially important that the 5′-terminal nucleotides should not be included within the helix, since the ribosome is preferentially attached to the 5′-end of the polynucleotide (see below). Then the relatively short helical regions of the polynucleotide will not present any serious hindrances to the work of the ribosome and the utilization of the entire nucleotide sequence of the polynucleotide in translation. In all probability, the working ribosome can uncoil short helical regions of the polynucleotide during the process of translation.

The absence of template activity in tRNA and the very low template activity of isolated ribosomal RNA are usually considered as a result of the large helical content of their chains or of the stability of their helical regions. However, since their helical content actually does not differ significantly from the helical content in single-stranded viral and messenger RNA's, it is most likely that the matter here may lie in a special stability of the helices. An important factor inhibiting template activity may also be the possible participation of the 5′-terminal nucleotides of the chain in helices. Finally, the presence of methyl groups and other "minor" nucleotides in tRNA and ribosomal RNA may also have a negative effect upon their template activity.

Evidently no principal limitations are imposed upon the primary structure of the template polynucleotide. It is known that not only natural mRNA's, but also varied artificial polyribonucleotides, containing nitrogen bases usually encountered in natural nucleic acids, can serve as templates (Nirenberg et al., 1963; Speyer et al., 1963). The nucleotide composition of polyribonucleotides may be quite varied, from homopolymers of the type of polyU, polyA, and polyC to random heteropolymers with the most varied proportions of A, G, U, and C. Only the homopolymer polyG and certain other synthetic polynucleotides with a predominance of G are very poor templates, but evidently mainly on account of the formation of very stable helices, i.e., secondary structure (Singer et al., 1963). Even polyribothymidylic acid can serve as a template (Szer, Ochoa, 1964).

Moreover, it is found that under special conditions, for example, in the presence of aminoglucoside antibiotics, not only polyribonucleotides, but also single-stranded polydeoxyribonucleotides can serve as templates for the work of the ribosome (McCarthy, Holland, 1965; Morgan et al., 1967). This means that even a difference in the sugar components of the two nucleic acids is not absolutely critical for the ability to interact with the ribosome and participate in protein synthesis. Nonetheless, in nature (in the cell) exclusively polyribonucleotides (mRNA) are evidently encountered as template polynucleotides.

Thus, in order for a polynucleotide to be able to serve as a template in a protein-synthesizing system, it should not possess a perfect secondary structure and must necessarily have nonhelical regions. The primary structure, however, may be quite varied, and the template polynucleotide can possess practically any nucleotide composition if its nitrogen components are the purine and pyrimidine bases usually encountered in natural nucleic acids.

3. Aminoacyl-tRNA

Aminoacyl-tRNA is the immediate form from which activated amino acid residues are transferred into the growing polypeptide chain in the ribosome. In other words, aminoacyl-tRNA's, just like the template polynucleotide, are components that interact directly with the ribosome during protein synthesis.

The formation of aminoacyl-tRNA from the free amino acids and tRNA's occurs outside the ribosome, entirely independent of it, in the "soluble" phase of the intraa

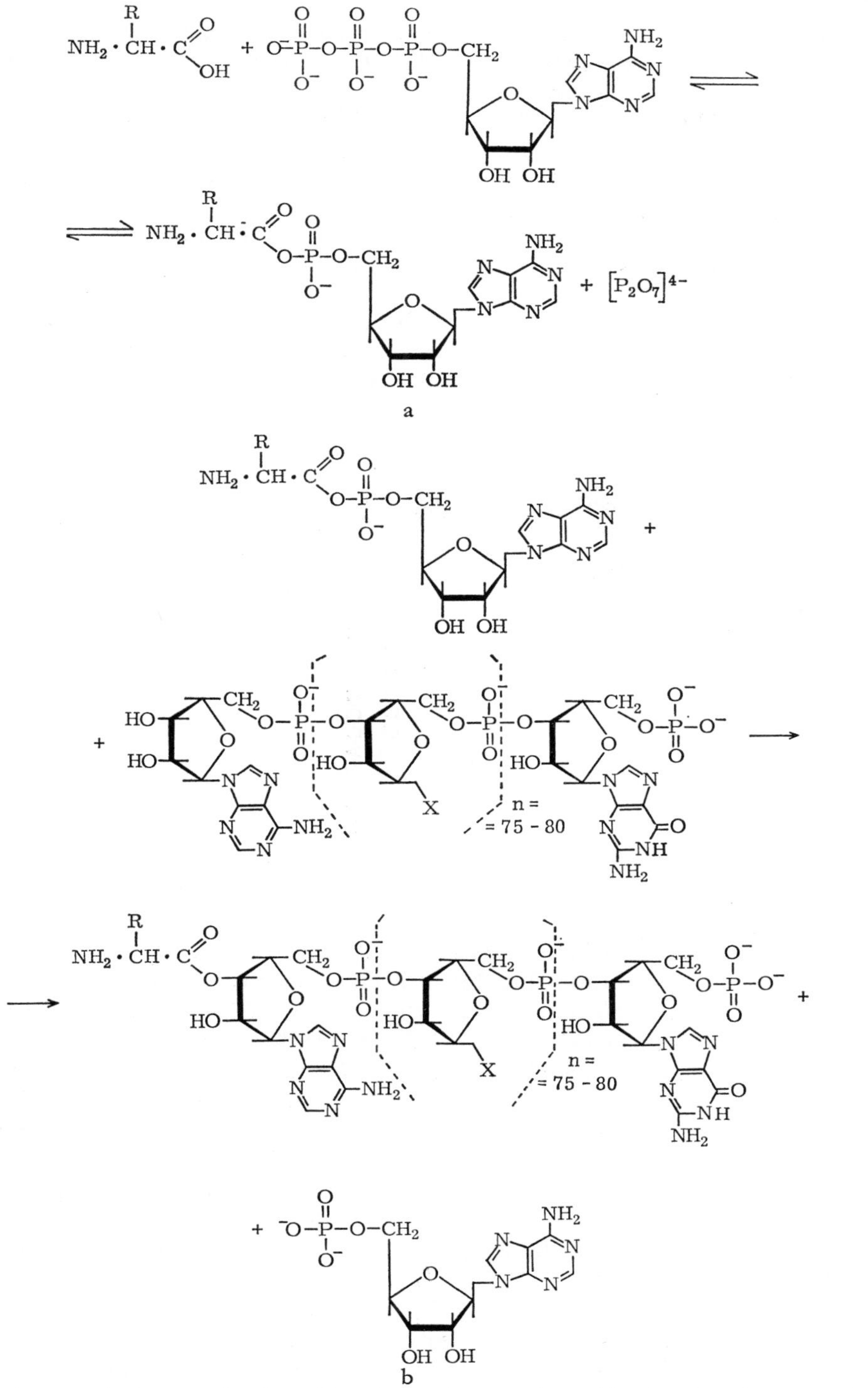

b

Fig. 16a and b. Activation of amino acid and accepting of amino acid residue by adaptor RNA

cellular medium or cell-free system (see survey: HOAGLAND, 1960). The process is catalyzed by the enzyme *aminoacyl-tRNA synthetase* (or pH 5 enzyme) and occurs in *two steps*. The first step is denoted as the reaction of *primary activation of carboxyl*: the free amino acid reacts with ATP, resulting in the production of aminoacyl adenylate and pyrophosphate (Fig. 16). The reaction product, aminoacyl adenylate, is bound to the enzyme in the form of a noncovalent complex. This aminoacyl adenylate – enzyme complex then reacts with free tRNA, and the second step takes place, denoted as the reaction of *acception* of the activated amino acid residue: the carboxyl of the amino acid residue of aminoacyl adenylate is transferred to the 3'-OH group of the ribose of

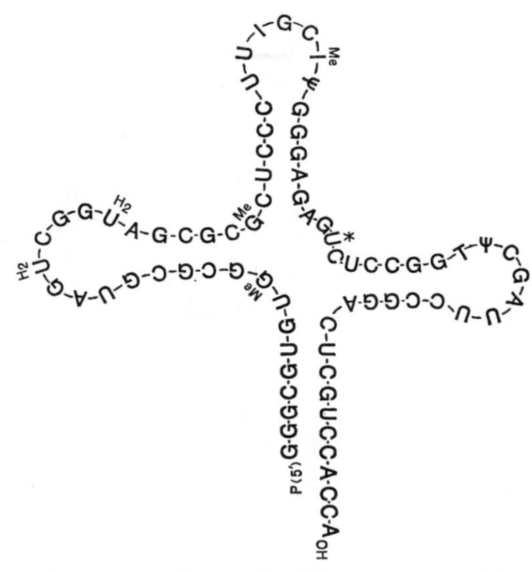

Fig. 17. Yeast alanine tRNA: complete nucleotide sequence and hypothetical clover leaf arrangement of the chain regions

the terminal adenosine of tRNA. As a result, AMP is released, and the final product, aminoacyl-tRNA, is formed (Fig. 16). It is important that each amino acid species has its own specific aminoacyl-tRNA synthetase, which activates only the given amino acid and then transfers it to tRNA. In turn, for each aminoacyl-tRNA synthetase there is its own specific tRNA (or tRNA's), and only the latter can interact with this enzyme in the reaction of its acception. Thus, as a result of the strict specificity of the enzyme with respect to the amino acid, on the one hand, and to tRNA, on the other, there is a specific acception of amino acids by quite definite tRNA's. According to this, alanine tRNA, serine tRNA, tyrosine tRNA, valine tRNA, phenylalanine tRNA, etc., can be distinguished. Henceforth they will be denoted as $tRNA_{ala}$, $tRNA_{ser}$, $tRNA_{tyr}$, $tRNA_{val}$, $tRNA_{phe}$, etc. The nucleotide sequence of yeast $tRNA_{ala}$ (HOLLEY et al., 1965) is given in Fig. 17.

Naturally, only aminoacylated tRNA's: alanyl-$tRNA_{ala}$, seryl-$tRNA_{ser}$, tyrosyl-$tRNA_{tyr}$, valyl-$tRNA_{val}$, phenylalanyl-$tRNA_{phe}$, etc., participate in the process of translation carried out by the ribosome. During protein synthesis with the participation of natural templates, i.e., mRNA, *the entire set of different aminoacyl-tRNA's* must be present in the protein-synthesizing system. When any of them is excluded, protein synthesis rapidly stops as a result of the inevitable breakoff of translation at the missing amino acid residue.

In the synthesis of polypeptides in cell-free systems with the participation of synthetic template polynucleotides, possessing an incomplete set of nitrogen bases, the presence of all the aminoacyl-tRNA's is not compulsory. Thus, for the process of translation to take place in the system with polyU, it is sufficient to supply it with phenylalanyl-tRNA alone (the triplet UUU codes phenylalanine — see Fig. 5). However, the presence of other aminoacyl-tRNA's in such a system may influence the rate of synthesis.

4. GTP

The functioning of the ribosome and the realization of the process of translation in a protein-synthesizing system requires the presence of guanosine-5′-triphosphate — GTP (KELLER, ZAMECNIK, 1956; NATHANS et al., 1962; CONWAY, LIPMANN, 1964). It cannot be replaced by any of the other known nucleoside triphosphates — ATP, UTP, CTP, etc. Guanosine-5′-diphosphate — GDP, on the contrary, is an inhibitor of translation. Thus, the GTP requirement of the system is very specific. In the course of translation GTP is cleaved into GDP and inorganic phosphate, and continuous regeneration of GDP to GTP is required.

5. "Transfer Factors" of Protein Nature

The interaction of the ribosome with the template polynucleotide and with aminoacyl-tRNA proves to be insufficient to ensure its functioning in the protein-synthesizing system. The presence in the system of certain special proteins, called transfer factors, is also essential (NATHANS, LIPMANN, 1961; NATHANS et al., 1962; NAKAMOTO et al., 1963; ALLENDE et al., 1964; etc.)[1]. These soluble proteins interact with the ribosome, and, being loosely associated with it, evidently participate in performing some necessary steps of the translation process.

At the present time, three soluble protein fractions which participate in the synthesis of the polypeptide chain on the ribosome as transfer factors are known in the bacterial protein-synthesizing system (LUCAS-LENARD, LIPMANN, 1966). The indicated proteins are not interchangeable, but must all be present together, i.e., they supplement one another. LIPMANN and associates denote them as the G-factor (GTP-dependent factor), the T_s-factor (the stable transfer factor) and the T_u-factor (the unstable transfer factor).

It should be indicated that in earlier studies of LIPMANN and co-workers (NAKAMOTO et al., 1963; ALLENDE et al., 1964), the transfer factors were divided into two fractions, which were denoted as A and B. In subsequent studies, as a result of the separation of the transfer factors by a different method, G and T fractions were obtained [NISHIZUKA, LIPMANN, 1966 (1, 2)]. The interrelations of these fractions, obtained by various methods, are as follows: $T = T_s + T_u$; $A = T_s$; $B = G + T_u$ (LUCAS-LENARD, LIPMANN, 1966).

[1] The need for special protein factors of the cellular sap for the functioning of the ribosomes was first demonstrated on animal systems (NATHANS, LIPMANN, 1960; VON EHRENSTEIN, LIPMANN, 1961; BISHOP, SCHWEET, 1961; FESSENDEN, MOLDAVE, 1961).

The G-factor. When this factor is added to complete ribosomes (70 S), GTPase activity appears, and GTP begins to be cleaved to GDP and inorganic phosphate. Neither the G-factor nor the ribosome individually is capable of cleaving GTP. Evidently a stoichiometric complex arises between the ribosomes and the G-factor, and GTPase activity is a characteristic only of this complex [NISHIZUKA, LIPMANN, 1966 (2); see also CONWAY, LIPMANN, 1964; CHAN, MCCORQUADALE, 1965; GORDON, LIPMANN, 1967).

Mg^{++} ions or substituting cations, Ca^{++} or Mn^{++}, are required for the association of the G-factor with the ribosome or for GTPase activity of the ribosome — G-factor system (GORDON, LIPMANN, 1967).

The cleavage of GTP after the addition of the G-factor to the ribosomes can occur in the absence of all other high-molecular weight components of the protein-synthesizing system and in the absence of protein synthesis [NISHIZUKA, LIPMANN, 1966 (2)].

The GTPase activity of the complete ribosome (70 S) with the G-factor is stimulated when the template polynucleotide is added [NISHIZUKA, LIPMANN, 1966 (2)]. It might be thought that a stimulating effect is exerted by binding of the ribosome to the template. However, no correlation is observed between the constant of binding of the polynucleotide to the ribosome and the stimulation of GTPase activity: for example, weakly bound polyA stimulates the cleavage of GTP twice as much as polyU. The addition of ribosomal RNA increases the GTPase activity of the ribosome — G-factor system only slightly, which is correlated with the very low template activity of this RNA.

On the other hand, the addition of tRNA also greatly stimulates GTPase activity of the ribosome — G-factor complex, regardless of whether the tRNA is amino-acylated or not [NISHIZUKA, LIPMANN, 1966 (2)]. Stimulation also occurs in the total absence of the template. It might be thought that stimulation in this case is due to the so-called "nonspecific" interaction (see below) of the 50 S subparticle of the ribosome with tRNA. The simultaneous presence of the template polynucleotide and tRNA in the system gives a substantial total effect of stimulation which somewhat exceeds the simple sum of the two effects separately.

The role of all the enumerated factors — combination of the 50 S with the 30 S, interaction with template polynucleotide, interaction with tRNA — in the increase of the GTPase activity of the ribosome with the G-factor is not clear. It may be that all these factors simply somewhat stabilize the structure of the ribosomal particle and, by restricting its flexibility, impart a more definite character to its conformation, as a result of which the ribosomes either react more completely with the G-factor or the structure of the created GTPase center itself becomes less variable.

The isolated G-factor, as a protein, is characterized by relatively great stability to freezing and thawing and to short heating to 50 °C (LUCAS-LENARD, LIPMANN, 1966). Association with the ribosome stabilizes it even more. The presence of mercaptoethanol or other SH compounds is required to maintain the activity of the G-factor in solution.

The system of the ribosome + G-factor is exceptionally specific with respect to the "substrate": it hydrolyzes only GTP and not any other known nucleoside triphosphates (ATP, UTP, CTP, etc.). GDP is a strong competitive inhibitor of the GTPase activity, and evidently the GTPase center possesses a great affinity for GDP (NISHIZUKA, LIPMANN, 1966 (2)].

The addition of GTP-γ-P^{32} and GTP-8-C^{14} to the system of ribosome + G-factor indicated that when GTP is cleaved, no transfer of the label to the ribosome is detected, i.e., neither labeled GDP nor labeled phosphate remains strongly bonded to it after the cleavage of GTP [NISHIZUKA, LIPMANN, 1966 (2)].

Thus, for functioning in a protein-synthesizing system, the ribosome must interact with a special soluble protein, denoted as the G-factor. As a result of such interaction, it acquires GTPase activity, essential for the functioning of the ribosome. The running of protein synthesis is not vital for the cleavage of GTP to GDP and phosphate; it can also occur in the absence of polypeptide synthesis. The addition of template polynucleotide and tRNA or aminoacyl-tRNA, however, substantially increases the GTPase activity of the system of ribosome + G-factor. The presence of SH compounds in the medium is essential for maintenance of the activity of the G-factor.

Since the G-factor combines with the ribosome and then participates in the cleavage of the high-energy bond of GTP, it is frequently stated that it may play a role in the supply of energy for "mechanical" displacement of the ribosome relative to the mRNA chain and for the simultaneous transfer of a tRNA residue from one site on the ribosome to another during the process of translation [CONWAY, LIPMANN, 1964; NISHIZUKA, LIPMANN, 1966 (2)]. Moreover, it might be thought that the decomposition of one molecule of GTP provides for the displacement of the ribosome by three nucleotide residues along the mRNA chain. Therefore, certain authors propose that this protein be called translocase. However, until the true role of the G-factor in translation has been strictly established, it seems more appropriate to refrain from using such a definite functional designation.

It should be mentioned that a protein fraction essential for the functioning of ribosomes and ensuring GTPase activity in the system has also been isolated from animal cells (ARLINGHAUS et al., 1964; FESSENDEN, MOLDAVE, 1963; IBUKI et al., 1966). The designation TF-I (transfer enzyme I), or "translocase" has been proposed for this GTP-dependent factor. The fraction TF-I was isolated by methods different from those used for the bacterial G-factor. It is not known to what degree the TF-I is related to the bacterial G-factor in properties and functions. In any case, there are substantial differences in properties here. The animal TF-I factor is extremely unstable in the isolated state, but it is stabilized by aminoacylated tRNA's. Free (deacylated) tRNA's do not exert any stabilizing effect. It is believed that TF-I forms complexes with aminoacyl-tRNA. SH compounds are not essential for the activity of TF-I (IBUKI et al., 1966).

The T_s-factor. The bacterial T_s-factor (LUCAS-LENARD, LIPMANN, 1966) coincides with the A fraction of transfer enzymes, characterized by Lipmann's laboratory in 1963–1964 (see, for example, ALLENDE et al., 1964). Without this protein factor in the system, there is no polypeptide synthesis in the ribosome, i.e., it is essential for the process of translation. In contrast to the G-factor, it does not directly participate in the step of the GTP cleavage during the process of translation.

The T_s-factor is even more stable to heating than the G-factor. It is also stable to freezing and thawing and to drying from the frozen state.

The T_s-factor is comparatively loosely bound to the ribosome. Thus, in the sedimentation of ribosomes through a layer of 10% sucrose, the T_s-factor is found to be entirely washed out from the ribosomes, although under these conditions the G-factor and especially the T_u-factor are not washed out. However, unwashed ribosomes retain a sufficient amount of the T_s-factor to ensure polypeptide synthesis in a cell-free system.

The T_u-factor. This factor, just like the T_s-factor, does not participate in the cleavage of GTP during translation, but it is essential for the maintenance of translation as a whole and for a polypeptide to be synthesized in the ribosome.

In physical and chemical properties it evidently represents a comparatively large and extremely unstable protein (Lucas-Lenard, Lipmann, 1966). In the isolated state in solution it is readily inactivated by freezing, by prolonged standing of the solution, and by heating. It is interesting that in the presence of other transfer factors (T_s and G), the T_u-factor is appreciably more stable. At the same time, serum albumin does not stabilize it. This may be an indication of some sort of specific interaction of the T_u-factor with the other transfer factors in solution [Nishizuka, Lipmann, 1966 (1)].

The T_u-factor is rather firmly bound to the ribosome. Sedimentation of the ribosomes through a layer of 10% sucrose does not remove it from the ribosome. It can be reliably washed out only by treating the ribosomes with strong salts — for example, 0.5 M NH_4Cl (in the presence of Mg^{++} to avoid unfolding) (Lucas-Lenard, Lipmann, 1966). The specificity of the interaction of the T_u-factor with the ribosome is indicated by the fact that, being bound to it, it acquires stability and in this form no longer loses its activity upon freezing and thawing, as well as upon prolonged storage in the frozen state (Lucas-Lenard, Lipmann, 1966).

The T-factors (T_s and T_u), just like the G-factor, require the presence of SH-compounds in solution to maintain their activity; however, whether they both require this, or whether only one of them is sensitive to oxidation cannot yet be said.

Recent reports indicate that T-factors take part in binding the GTP and aminoacyl-tRNA with the ribosome (Allende et al., 1967; Lipmann et al. 1967; Gordon, 1967).

A transfer fraction essential for the work of the ribosomes, but not participating in GTP cleavage, has also been isolated from animal cells; it has been called the TF-II fraction (transfer enzyme II), or "peptide synthetase" (Arlinghaus et al., 1964; Fessenden, Moldave, 1961, 1962; Sutter, Moldave, 1966). TF-II combines with the ribosome, and this protects it from inactivation by glutathione, to which it is subject in the free state. Nonetheless, the relationship of the animal TF-II fraction to the bacterial T-factors is unclear.

6. Protein "Initiation Factors"

It was noted that in cell-free protein-synthesizing systems with thoroughly washed ribosomes, a reading of natural mRNA's is hindered, despite the presence of all the components. In unpurified systems, there is no such hindrance. These observations indicated that the reading of natural mRNA's requires not only the known components of the protein-synthesizing system, but also some additional factors, rather loosely bound to the ribosomes, and washed from them during careful purification. These factors have been found to be essential only for the *beginning* of the translation process. These factors evidently are not required for the course of translation itself.

The indicated "initiation factors" proved to be substances of a protein nature. They were detected and identified in 1966 simultaneously in three laboratories (REVEL, GROS, 1966; BRAWERMAN, EISENSTADT, 1966; STANLEY et al., 1966).

According to the data of Ochoa's laboratory [STANLEY et al., 1966; SALAS et al., 1967 (1, 2)], there are at least two protein "initiation factors", F1 and F2 (for the system of *E. coli*), which provide for the initiation of translation of natural mRNA, including viral RNA's, and only those specific polynucleotides that possess special initiation codons at the 5′-end, AUG or GUG. They do not influence the initiation of translation of other synthetic polynucleotides. Evidently, the action of these two factors is that when being associated with the ribosome they stimulate its initial association with the initiator tRNA (formylmethionyl-tRNA) (see below) in the presence of the initial nucleotide combination of the mRNA chain, AUG or GUG.

The "initiation factor" of another nature was isolated in Gros' laboratory (REVEL, GROS, 1966, 1967; GROS, 1967). In contrast to the "initiation factors" F1 and F2 (denoted by different symbols), the initiation factor isolated by REVEL and GROS stimulates the attachment of the ribosome to the 5′-end of the template polynucleotide, i.e., acts prior to the binding of the initiator formylmethionyl-tRNA.

7. "Termination Factors"

Besides the proteins, necessary for the initiation and continuation of translationl the complete protein-synthesizing system must also contain, as was found, specia, macromolecular components responsible for the release of the completed polypeptide chain from the ribosome after the termination of translation (GANOZA, 1966). A soluble "releasing factor" of a protein nature was recently isolated from *E. coli* (CAPECCHI, 1967). It was shown that in order to exhibit its activity, the "releasing factor" must come into an interaction with the ribosome — mRNA — peptidyl-tRNA complex.

8. Inorganic Cations

The work of the protein-synthesizing system primarily requires a definite concentration of divalent cations, generally of Mg^{++} ions. Usually in a cell-free system the optimum concentration of Mg^{++} ions is from 0.005 M to 0.015 – 0.02 M. In the presence of increased concentrations of monovalent cations (K^+, NH_4^+, tris$^+$) in the medium, somewhat larger concentrations of Mg^{++} are required than in the presence of a low content of monovalent cations. When working with synthetic template polynucleotides, as a rule, an increased Mg^{++} concentration is required for the beginning of translation, 0.014 – 0.016 M and above; after translation has begun, the Mg^{++} concentration may be lowered, for example, to 0.008 M. With natural mRNA, the initiation of translation occurs successfully at the same Mg^{++} concentration that is optimum for the entire process of translation, for example, 0.008 M (however, as was noted above, when natural mRNA's are used, the presence of special proteins is required for the initiation of translation). Evidently, at the given concentration of Mg^{++} ions, a whole series of essential interactions in the protein-synthesizing system is ensured: the combination of the ribosome with transfer factors and the association with mRNA and tRNA. It may be that the same concentrations are optimum for the work itself of the ribosome and of transfer factors in polypeptide synthesis.

Although Mg^{++} usually plays the role of a predominant and necessary divalent cation in the protein-synthesizing system *in vivo*, in systems *in vitro* it can be replaced with more or less success by Ca^{++}, as well as by Mn^{++}, in suitable concentrations (0.01 M) in the medium (MOORE, 1966; GORDON, LIPMANN, 1967).

Either K^+ or NH_4^+ is required as another essential cation for the protein-synthesizing system (LUBIN, 1963, 1964; LUBIN, ENNIS, 1963, 1964; CONWAY, 1964; SPYRIDES, 1964). These two cations are apparently interchangeable. Their optimum concentration in the medium is evidently of the order of 0.1 M. Na^+ and Li^+ not only cannot replace K^+ or NH_4^+, but are antagonists of them, inhibiting the work of the protein-synthesizing system. The presence of K^+ or NH_4^+ in the concentrations indicated above is apparently especially vital during the initiation of polypeptide synthesis; subsequently, if peptide synthesis has already begun, their concentration can be substantially lowered without any appreciable effect upon the process of translation. In exactly the same way, the addition of Li^+ after the beginning of peptide synthesis has no significant inhibiting effect upon this synthesis, although Li^+, added to the system beforehand, inhibits the synthesis (LUBIN, 1963). Perhaps, K^+ and NH_4^+, as specific cations, are needed to create some sort of conditions on the ribosome providing for the formation of the initial complex of ribosome — template — aminoacyl-tRNA and for the initiation of translation (see below). However, in all probability they are also required for the reaction of synthesis of the peptide bond itself in the ribosome (TRAUT, MONRO, 1964; SMITH et al., 1965; RYCHLIK, 1966; RYCHLIK et al., 1967).

II. Association of the Ribosome with Components of the Protein-Synthesizing System

The ribosome carries out its function of protein synthesis in association with other high-molecular weight components of the protein-synthesizing system. Transfer factors G and T and three initiation factors F evidently represent an integral constituent of the functioning ribosome and may be present in it in a permanent, although labile, association. But the template polynucleotide, and all the more, aminoacyl-tRNA always represent components only *temporarily associated* with the ribosome; during the process of functioning, the ribosome necessarily changes templates, and even with one template there is a succession of tens and hundreds of aminoacyl-tRNA's. Therefore the association of the ribosome with the *template polynucleotide* and with *aminoacyl-tRNA* is of special interest as a manifestation of definite aspects of the functioning of the ribosome.

1. Association of the Ribosome with the Template Polynucleotide

It can be shown in experiments *in vitro* that purified bacterial ribosomes can associate with certain polynucleotides in the absence of all other high-molecular weight components of the protein-synthesizing system.

a) Conditions of Complex Formation and Stability of the Complex

The presence in the medium of only a definite concentration of Mg^{++} ions, as a rule, no lower than 0.005 M, can be sufficient for the formation of a more or less stable complex of the ribosome with the template polynucleotide. Under such conditions, i.e., without any other high-molecular weight components of the protein-synthesizing system, in the presence of 0.005 to 0.015 M Mg^{++}, complex formation of the *E. coli* ribosomes with polyU [SPYRIDES, LIPMANN, 1962; TAKANAMI, OKAMOTO, 1963 (1); MOORE, 1966 (1)], polyC [HASELKORN, FRIED, 1964; MOORE, 1966 (1)], yellow turnip mosaic virus RNA (HASELKORN et al., 1963; VOORMA et al., 1964), polio virus RNA (WARNER et al., 1963), RNA from f2 virus (TAKANAMI et al., 1965), as well as with polyribothymidylic acid (SZER, NOWAK, 1966), with single-stranded and denatured DNA [TAKANAMI, OKAMOTO, 1963 (2)], and even with derivatives of polyU with a lack of template activity — poly-N-methyluridylic acid and poly-5,6-dihydrouridylic acid (SZER, NOWAK, 1966) have been reported.

In any case, such association is evidently *reversible* in the sense that there is a dynamic equilibrium between the complex, on the one hand, and the free ribosomes and free templates on the other:

$$\text{Ribosome} + \text{template} \rightleftharpoons \text{ribosome} - \text{template}.$$

The bound template of the complex therefore can be comparatively readily exchanged for the exogenous free template. At the same time, in the presence of 0.005 to 0.015 M Mg^{++}, the constants of association of the template polynucleotides enumerated above with the ribosomes are rather great, while the rate of dissociation is comparatively low, so that the corresponding complexes can be registered even when the excess free templates are washed out.

Among the templates enumerated above, important differences are observed in the constants of their binding to the ribosomes. Thus, complexes of the ribosomes with polyU are appreciably stronger than those with viral RNA's and all the more than those with polyC [HASELKORN, FRIED, 1964; MOORE, 1966 (1); DAHLBERG, HASELKORN, 1967]. Therefore, polyU can displace polyC and RNA from a complex with the ribosomes. Appreciably stable complexes of the ribosomes with polyA are not formed at all under the indicated conditions [TAKANAMI, OKAMOTO, 1963 (1); MCLAUGHLIN et al., 1966].

When the Mg^{++} ion concentration is lowered to 10^{-3} or 10^{-4} M, the stability of the complex between the ribosome and the template polynucleotide is greatly lowered, the rate of dissociation becomes high, and therefore the already formed ribosome — mRNA or ribosome — polyU complexes rapidly and completely dissociate into their components, the ribosome and the polynucleotide, during washing to remove the free components. However, in an equilibrium system the ribosome — template complex is detected even at 10^{-4} M Mg^{++} (DAHLBERG, HASELKORN, 1967).

There is an indication that for the association of the ribosome with the template polynucleotide (polyU), Mg^{++} ions in the medium can be successfully replaced by Ca^{++} or Mn^{++} in the same concentration, about 10^{-2} M, as well as by 10^{-3} M spermidine [Moore, 1966 (1)]. This means that Mg^{++} is not very specific in the association of the ribosome with the template polynucleotide. K^{+} ions somewhat destabilize the complex formed with the participation of divalent cations (Dahlberg, Haselkorn, 1967).

Temperature conditions, in any case within the range from 0° to 40°C, evidently are not critical for the formation of complexes of the ribosomes with polyU, polyC, and viral RNA's. The complex can be formed both at 0 – 4°C and at 37° [Spyrides, Lipmann, 1962; Takanami, Okamoto, 1963 (1); Haselkorn et al., 1963; Haselkorn, Fried, 1964; Warner et al., 1963; Takanami et al., 1965]. It has been shown, however, that when the temperature is increased, the strength of the complex (binding constant) is appreciably lowered (Dahlberg, Haselkorn, 1967). At the same time, it has been noted in certain cases that the ribosomes can associate better with the template at increased temperatures (within the permissible temperature range). For example, a preparation of polio virus RNA adds more ribosomes at 37° than at 4°C (Warner et al., 1963). This peculiarity may be due not to a direct effect of temperature upon the forces of association between the ribosome and mRNA, but to the fact that the secondary structure of mRNA may prevent its association with the ribosome, while at an increased temperature the secondary structure is partially broken down or rendered more labile.

b) Region of the Polynucleotide Interacting with the Ribosome

As a rule, the chain of the template polynucleotide is significantly longer than the linear dimensions of the ribosomal particle. Therefore, the ribosome can associate only with some restricted region of the polynucleotide. The size of this region is unknown. For the case of association of the ribosome with polyU, there has been an attempt to estimate its size by determining the length of the section of the polynucleotide that is protected from the action of exogenous ribonuclease — as is believed, due to binding with the ribosome. For this purpose, polyU was associated with ribosomes and then subjected to the action of ribonuclease. After deproteinization of the hydrolyzate, polynucleotide (polyU) fragments about 25 nucleotide residues long were found (Takanami, Zubay, 1964). From this it was concluded that a polyU region, about 25 nucleotide residues long (up to 100 to 150 Å in length), is attached to the ribosome.

Evidently the region to which the ribosome can be attached in no case can be double stranded [Takanami, Okamoto, 1963 (1, 2)]. It should be assumed that the ribosome can associate only with a *single-stranded nonhelical region* of the template polynucleotide.

During the association with the template polynucleotide, the ribosome probably exhibits a special affinity for the 5′-end of the polynucleotide. It can be shown either in experiments with the association *in vitro* (Matthaei et al., 1964) and on the basis of analysis of data on reading out the templates *in vivo* (Kepes, 1967; Janofsky, Ito, 1966; Martin et al., 1966).

c) Ribosomal Site Interacting with the Template Polynucleotide

Any consideration of the question of the site on the ribosome that interacts with the template polynucleotide is based upon the following most important observation: each ribosome is capable of binding only one chain of the template polynucleotide [MOORE, 1966 (1)]. Consequently, there must be only *one* functional site on the ribosome, interacting with the template polynucleotide. (This does not mean that there is necessarily only one point of interaction or one center of such interaction on the ribosome. The polynucleotide may be retained on the ribosome by several binding centers, but since they cannot in any way play the role of independent binding sites, we should consider them totally as a single binding cooperative.)

It cannot be said beforehand whether the mRNA binding site is localized exclusively on one of the two subparticles of the ribosome (50 S or 30 S) or whether it is formed only by the large and small subparticles jointly. This question can be resolved experimentally in the following way: the ribosome is dissociated into the larger and smaller subparticles, they are separated, and template polynucleotide and Mg^{++} to a concentration of 10^{-2} M are added to each of them individually. It is found that under these conditions the 30 S subparticle binds the template, while the 50 S subparticle does not bind it [TAKANAMI, OKAMOTO, 1963 (1); OKAMOTO, TAKANAMI, 1963; DAHLBERG, HASELKORN, 1967]. On this basis, it is now believed that the mRNA binding site of the ribosome is localized *on the smaller, 30 S, subparticle of the ribosome.* Both indirect experimental data [MOORE, 1966 (2)] and a consideration of the mechanism of the functioning of the ribosome during translation (Chapter III, Section 2), permit us to think that the mRNA binding site is situated on the portion of the 30 S subparticle that adjoins the 50 S subparticle of the ribosome (Fig. 18).

There are definite experimental indications that the mRNA binding center of the ribosome is formed chiefly by the *protein component* of the 30 S subparticle. Thus, after very mild treatment of the 70 S ribosome or its 30 S subparticle with trypsin, their ability to bind the template polynucleotide is reduced or lost, despite the preservation of apparent integrity of the particles [ZAK et al., 1966; KAJI et al., 1966 (1)]. More direct indications were recently obtained by NOMURA and TRAUB (NOMURA, TRAUB, 1966). They showed that the separation of a definite portion of the ribosomal protein from the 30 S subparticle deprives it of its ability to bind the template polynucleotide (poly U). If the protein portion removed is separated into two fractions, then it is found that the addition of only one of them to the "stripped" 30 S subparticles restores their ability to bind poly U. This fraction evidently includes no more than four to six protein molecules per 30 S subparticle. From this it might be thought that the mRNA-binding ability of the ribosome can be provided for by a small special group of ribosomal proteins. It is likely that this protein mRNA-binding center of the 30 S subparticle may interact only with a very limited section of the template polynucleotide — for example, down to the size of one triplet of nucleotides. The amino groups of the ribosomal protein evidently do not participate in the binding of the template polynucleotide [MOORE, 1966 (2)].

It is not excluded that in addition to the binding of a small section of the template at the mRNA-binding center of the ribosome, there may also be a possibility of complementary interactions between other sections of the template polynucleotide and the ribosomal RNA. There are experimental indications in support of such a possibility. For example, it has been shown that poly U, and also, to a lesser degree, poly C, in the presence of 0.01 M Mg^{++}, can interact complementarily with certain regions of isolated ribosomal RNA in solution (Moore, Asano, 1966; see also Millar et al., 1965). Poly A does not bind with ribosomal RNA. Furthermore, direct data were obtained in support of the participation of the amino groups of ribosomal RNA in the binding of poly U with the ribosome [Moore, 1966 (2)]. From this it may be assumed that the ribosome contains single-stranded purine-rich sequences (blocks) of ribosomal RNA, which can complementarily interact with sections of the template polynucleotide situated outside the protein mRNA-binding center. This should provide those *supplementary* retaining forces that are observed in the case of the association of the ribosome with polypyrimidine polynucleotides. In other words, the following hypothesis can be formulated. All the template polynucleotides interact with the protein mRNA-binding center on the 30 S subparticle, but the strength of this interaction within the pure system ribosome + template is not so great as to ensure the formation of a stable complex. The presence of polypyrimidine sequences in the template polynucleotides provides them with the possibility of complementary interaction with the polypurine regions of the ribosomal RNA which provides an additional contribution to the stability of the ribosome — template complex. It may be appropriate to mention that poly U binds better to the ribosome than poly C, possibly solely because the latter has a tendency to form its own secondary structure which competes with the indicated additional complementary binding.

Another hypothesis on the binding of the template polynucleotide to the ribosome is advanced by Moore [Moore, 1966 (2)]. On the basis of his own experimental data on the participation of the amino groups of ribosomal RNA and indirect conclusions of the non-participation of the amino groups of the template polynucleotide in binding to the ribosome, he presumes that the template polynucleotide is retained by hydrogen bonds between its phosphates, on the one hand, and the amino groups of ribosomal RNA, on the other.

Two most probable hypotheses can be advanced on the role of Mg^{++} ions or divalent cations replacing them in the binding of mRNA to the ribosome. First of all, in principle they can participate directly in the formation of bonds between the ribosomal particle and the template. For example, they might form Mg^{++}-bridges between the phosphates of the template polynucleotide, on the one hand, and the carboxyls of the ribosomal protein and phosphates of the ribosomal RNA, on the other. However, the vital role of Mg^{++} ions in the association of the ribosome with the template polynucleotide might also be indirect and consist chiefly of providing the necessary shielding of the negatively charged phosphates of the ribosome and template polynucleotide, so as to suppress their electrostatic repulsion. It is most probable that both mechanisms take place, i.e., the association of the template to the ribosome is stabilized both on account of nonspecific neutralization of negative charges and on account of direct participation of divalent cations (Dahlberg, Haselkorn, 1967).

d) Summary

The complex formation of purified ribosomes with template polynucleotides in the absence of all other high-molecular weight components of the protein-synthesizing system requires the presence of divalent cations, mainly Mg^{++}; Ca^{++} and Mn^{++} can substitute it. At a Mg^{++} concentration in the medium no lower than 0.005 M (from 0.005 to 0.015 M), the ribosomes form a stable complex with polyU, less stable with certain natural mRNA's, still less stable with polyC, and entirely unstable with polyA. The ribosome can bind only to the single-stranded nonhelical regions of the polynucleotides apparently in the main with the 5′-end of the chain. Each ribosome can bind only one polynucleotide. The 30 S subparticle of the ribosome is responsible for binding with the template polynucleotide. Probably there is a special mRNA binding center on the 30 S subparticle, formed chiefly by ribosomal protein; it may be that this site interacts only with a small section of the template polynucleotide, down to the size of one triplet of nucleotides. The possibility remains that the additional forces binding the template polynucleotide with the ribosome can be provided for by complementary interactions of the polypyrimidine sections of the template with nonhelical polypurine regions of the ribosomal RNA. The role of divalent cations (Mg^{++}) may consist both of the formation of ion bridges between the electronegative groups of the polynucleotide and the ribosome and simply of nonspecific suppression of the electrostatic repulsion of the charged groups.

2. Formation of a Ternary Complex Ribosome — Template — Aminoacyl-tRNA

It was shown above that pure ribosomes *in vitro* can associate with certain polynucleotides, for example, polyU, giving a more or less stable binary complex of ribosome and template. However, for other template polynucleotides, for example, polyA, such an association proves to be extremely labile, i.e., in general a sufficiently firm complex is not formed. Nonetheless, in the presence of all the components of the protein-synthesizing system, polypeptide synthesis proceeds with equal success in both cases. It can be shown that the substantial stability of the ribosome–template complex is imparted by the participation of aminoacyl-tRNA in it (Hatfield, 1965, 1966; McLaughlin et al., 1966). Since *in vivo* the formation of ribosome–template complexes obviously occurs not in a pure binary system, but in the presence of a whole set of components of the protein-synthesizing apparatus, evidently association in a ternary complex of ribosome–template–aminoacyl-tRNA better approximates the natural interrelationships between ribosome, mRNA, and tRNA. The binary association of the ribosome with the template polynucleotide can evidently have a real meaning as an intermediate (though even unstable) step in the formation of the ternary complex:

$$\text{ribosome} + \text{template} + \text{aa-tRNA} \rightleftharpoons [\text{ribosome–template} + \text{aa-tRNA}] \rightarrow$$
$$\rightarrow \text{ribosome–template–aa-tRNA}.$$

a) Conditions of Formation of the Ternary Complex and Its Stability

Hence, if we have the equilibrium system considered above,

$$\text{ribosome} + \text{template} \xrightleftharpoons{10^{-2}\text{ M Mg}} \text{ribosome-template complex},$$

the introduction of aminoacyl-tRNA into this system leads to the formation of a relatively stable ribosome-template-aminoacyl-tRNA complex [Kaji, Kaji, 1963, 1964 (1); Nakamoto et al., 1963; Spyrides, 1964; Nierenberg, Leder, 1964; Leder, Nierenberg, 1964 (1, 2)]. Only the aminoacyl-tRNA *that corresponds to the codons* of the template polynucleotide enters into this complex, i. e., there is *a strictly specific selection of aminoacyl-tRNA's* from the medium.

A rather stable ternary complex of ribosome–template–aminoacyl-tRNA is formed to an equal degree when template polynucleotides characterized by a high constant of binding to the ribosome (for example polyU) or a low binding constant (polyA or oligonucleotides) are used [NIRENBERG, LEDER, 1964; KAJI, KAJI, 1964 (1)], i.e., the position of equilibrium of the reaction cited above evidently is of no great significance for the result of formation of the ternary complex. When aminoacyl-tRNA is introduced into the complex, the substantial difference in the strength of the binding between the ribosome and polynucleotides of different compositions, indicated above, disappears. For example, the introduction of lysyl-tRNA into the system of ribosome + polyA makes the binding of the ribosome to polyA even somewhat stronger than the binding of the ribosome to polyU, in the presence or in the absence of phenylalanyl-tRNA (MCLAUGHLIN et al., 1966). Thus, the introduction of aminoacyl-tRNA actually ensures the formation of a stable complex, independent of the intrinsic affinity of the ribosome for the template polynucleotide.

Evidently, for the same reasons, the ribosome in the presence of the corresponding aminoacyl-tRNA's can form rather stable complexes with quite short oligonucleotides — for example, trinucleotides (NIRENBERG, LEDER, 1964), which themselves do not practically bind to the ribosome (HATFIELD, 1965, 1966).

It is important to note that the participation of all three components — the ribosome, template, and aminoacyl-tRNA — in the formation of a stable complex is cooperative. For example, neither phenylalanyl-tRNA nor the trinucleotide UUU or UUC appreciably bind to the ribosome individually, but the addition of UUU or UUC stimulates the binding of phenylalanyl-tRNA; the addition of phenylalanyl-tRNA stimulates the binding of UUU or UUC, and the increase in the binding of the trinucleotides in this case is approximately equal to the increase in the binding of the aminoacyl-tRNA (HATFIELD, 1965, 1966). The same conclusions can be drawn in analyzing the binding of polyA and lysyl-tRNA to the ribosome (MCLAUGHLIN et al., 1966).

The formation of ternary complexes requires the same concentrations of Mg^{++} ions as the binding of the ribosome to polyU, polyC, or viral RNA in the absence of aminoacyl-tRNA — 0.005 to 0.015 M. However, to obtain stable complexes with the participation of trinucleotides, higher concentrations of Mg^{++} ions are often used — 0.02 to 0.03 M (NIRENBERG, LEDER, 1964).

From a number of reports it is evident that the formation of the ternary complex is specifically stimulated by K^+ or NH_4^+ cations (in concentrations up to 0.05 to 0.1 M) — according to some data strongly (SPYRIDES, 1964), and according to other only slightly [PESTKA, NIRENBERG, 1966 (1)]. On the contrary, the presence of Na^+ or Li^+ ions can inhibit the formation of a complex, according to some authors [SPYRIDES, 1964; KAJI, KAJI, 1964 (1, 2)] strongly, and according to others negligibly [PESTKA, NIRENBERG, 1966 (1)].

Evidently no other components of the protein-synthesizing system are required for the formation of the ribosome–template–aminoacyl-tRNA ternary complex, in any case at the Mg^{++} concentration of 0.01 M and higher [SPYRIDES, LIPMANN, 1962; CONWAY, 1964; KAJI, KAJI, 1964 (1); SPYRIDES, 1964; NIRENBERG, LEDER, 1964; NISHIZUKA, LIPMANN, 1966; KURLAND, 1966]. It is known, however, that at concentrations of Mg^{++} closer to the "physiological", of about 0.005 M, the specific binding of aminoacyl-tRNA with the ribosome also needs the presence of protein T-factors and GTP.

The complex formed retains its stability within a rather broad range of temperatures — from 0° up to at least 45–50 °C in the presence of 0.03 M Mg^{++}, independent of the composition of the template polynucleotide (McLaughlin et al., 1966). When the temperature is further increased, the complex dissociates; the dissociation is quite reversible, so that the complex is reformed upon cooling. It has been proposed that the temperature at which half of the complexes in the system are dissociated be called the "temperature of melting" and that it be used to characterize the stability of the complex (McLaughlin et al., 1966). The "temperatures of melting" of ternary complexes containing various template polynucleotides in the presence of 0.03 M Mg^{++} in the medium are:

54–56° — for the complex of ribosomes of *E. coli* with polyU + phenylalanyl-tRNA,

60° — for the complex of ribosomes with polyA + lysyl-tRNA,

58° — for the complex of ribosomes with poly(U, A) + isoleucyl-tRNA[1],

53° — for the complex of ribosomes with poly(U, A) + tyrosyl-tRNA[1].

The differences observed, depending upon the composition of the template polynucleotide and nature of the aminoacyl-tRNA, are quite reproducible, but, as can be seen, they are relatively small, i.e., all these complexes are rather close in stability.

The components of the ternary complex are practically not exchanged with exogenous components in the cold in the presence of 0.02–0.03 M Mg^{++}, at any rate, no visible exchange of aminoacyl-tRNA can be noted during a period of 10–20 min (Nirenberg, Leder, 1964; McLaughlin et al., 1966). Evidently the complex is extremely stable under these conditions, the association constant is very great, and the rate of dissociation is low. At higher temperatures, for example 24 °C, the exchange of aminoacyl-tRNA, bound in the complex ribosome–triplet–aminoacyl-tRNA, for exogenous aminoacyl-tRNA of the same specificity is also very slow (Hatfield, 1966). However, the template triplet of this complex quickly exchanges for an identical exogenous triplet (Hatfield, 1966). The possibility remains that under conditions approaching the physiological, 30° or 37°, 0.01 M Mg^{++}, an appreciable exchange of the components of the complex, especially mRNA, may take place with the exogenous components.

It is important to emphasize that reversible and equilibrium association may be characteristic only of the stage of formation of the nontranslating complex. The subsequent initiation of translation (see below, Section III, 1), i.e., the formation of the *translating* complex, firmly reinforces the association of the ribosome with the template polynucleotide and makes it irreversible.

Thus, a more or less stable association of the ribosome with the template polynucleotide requires the presence of a definite concentration of Mg^{++} ions in the medium (0.005 to 0.015 M), and the corresponding aminoacyl-tRNA must take part

[1] AUU codes isoleucine, UAU codes tyrosine (see Fig. 5).

in the association so that a ternary complex of ribosome–template–aminoacyl-tRNA is formed. The process of association does not require external energy, i.e., represents a thermodynamically and kinetically spontaneous reaction.

b) Region of the Template Polynucleotide Associating with the Ribosome and tRNA

All that has been said of the region of the template polynucleotide associating with the ribosome, forming a binary ribosome–template complex evidently also remains in force for the case of the ternary ribosome–template–aminoacyl-tRNA complex: this region must be necessarily single-stranded, but not double helical. In this case, this is especially important, since the associating region of the template polynucleotide contains a *codon*, the nitrogen bases of which should be open for complementary interaction with the anticodon of the aminoacyl-tRNA.

The experiments of NIRENBERG and LEDER (NIRENBERG, LEDER, 1964) have shown that as little as the trinucleotide associates with the ribosome and aminoacyl-tRNA corresponding to it, giving a ternary complex. The properties of such a complex do not differ essentially from the properties of a complex in which any long polynucleotide participates. Moreover, a determination of the thermal stability of the ternary complexes containing polyA and oligoA of various lengths (ribosome–polyA–lysyl-tRNA) gave the following result: the stability of the complex in which the trinucleotide $(pA)_3$ participates practically does not differ from the stability of a complex containing polyA more than 100 nucleotides long; the same stability is given by oligoA six or eight nucleotides long (MCLAUGHLIN et al., 1966). This permits us to believe that as little as the trinucleotide section of the polynucleotide, about 10 Å long, is already sufficient for binding with the ribosome and tRNA in a ternary complex. In such a case, it is precisely the *short double-helical codon-anticodon pair* that must evidently be the *basic point of retention* of the template polynucleotide (just like tRNA — see below) on the ribosome in the ternary ribosome–template–aminoacyl-tRNA complex.

However, the thermal stability of the complex containing polyU (ribosome–polyU–phenylalanyl-tRNA) depends upon the length of the polynucleotide: polyU with a length of over 100 nucleotides gives appreciably higher stability than oligoU eight nucleotides long, and an even higher stability than oligoU with a length of four nucleotides (MCLAUGHLIN et al., 1966). This may be due to the circumstance already noted, that polypyrimidine polynucleotides may complementarily interact with ribosomal RNA through other sections of their chain, thereby providing an additional contribution to the stability of the complex.

In studies of the formation of ternary complexes containing oligonucleotides (U_3, U_4, A_3, etc.) as the template, it was noted that with oligonucleotides without a 5′-terminal phosphate complex formation is poorer than in the case of oligonucleotides with a phosphorylated 5′-end, while the presence of phosphate on the 3′-end, on the contrary, inhibits complex formation (NIRENBERG, LEDER, 1964). Methylation of the terminal 5′-phosphate also inhibits complex formation (ROTTMAN, NIRENBERG, 1966). Evidently the phosphate residue at the 5′-end of the template polynucleotide somehow promotes association of the ribosome with the polynucleotide and tRNA. Therefore it might be thought that the phosphorylated 5′-end region of the template polynucleotide possesses *an advantage over the internal regions of the polynucleotide* in the formation of a ternary ribosome–template–aminoacyl-tRNA complex.

c) Regions of Aminoacyl-tRNA Interacting with the Template Polynucleotide and the Ribosome

As has already been stated, the ternary complex of ribosome–template–aminoacyl-tRNA strictly selectively includes only the aminoacyl-tRNA that bears the amino acid coded by the template. At the same time, according to the adaptor hypothesis (CRICK, 1958; CHAPEVILLE et al., 1962), the specificity of the aminoacyl residue itself should not play a role in the ensurance of selectivity of binding of aminoacyl-tRNA in the complex. Moreover, the experimental data show that the aminoacyl residue generally is not necessary for specific binding of tRNA in the ternary complex: a free (non-acylated) tRNA, specific for the amino acid coded by the template, is also well bound with the ribosome in the presence of the template [KAJI, KAJI, 1964 (1); KURLAND, 1966; LEVIN, 1966; SEEDS and CONWAY, 1966]. Consequently, the aminoacyl residue of the aminoacyl-tRNA molecule evidently does not participate, or participates little in the binding and the more so, does not determine the code specificity of the binding.

It is clear that if the aminoacyl residue is not vital for selective binding of the aminoacyl-tRNA, then only the polynucleotide chain of the tRNA must interact with the codon of the template polynucleotide, determining the selectivity of the binding. This interaction should be strictly specific, i.e., evidently it must be a *complementary* interaction of the codon with a definite region of the tRNA chain. Consequently, each tRNA molecule must have a special region, complementary with the codon that codes the amino acid accepted by the given tRNA. This region, consisting of three nulceotide residues, has been denoted as the *anticodon*. The anticodon occupies approximately the middle position in the tRNA chain (from the 34 — 36th to the 36 — 38th nucleotide). All the selectivity of the binding of a given specific aminoacyl-tRNA in the presence of the corresponding codon in the ribosome is determined by the nucleotide composition and sequence of the anticodon region of tRNA.

It may be assumed that the complementary binding of the anticodon of tRNA with the codon of the template polynucleotide is also quantitatively the basic force of retention of the aminoacyl-tRNA in the complex. Of course, under conditions of an aqueous solution, the complementary interaction of only three nucleotides usually is very weak and unquestionably could not provide for a sufficiently strong retention of the large aminoacyl-tRNA molecule on the template polynucleotide (LIPSETT et al., 1961; LIPSETT, 1964; MICHELSON, 1965; MCLAUGHLIN et al., 1966). However, in suitable surroundings on the ribosome, the stability of the codon-anticodon interaction can be substantially higher than in solution.

Although there is almost no doubt that the anticodon plays a most important role in ensuring the binding of the aminoacyl-tRNA in a ternary complex, a possible role of other regions of tRNA in its retention on the ribosome also cannot be excluded.

Firstly, a supplementary, even though a weak binding of tRNA by some other regions of its chain, if it occurred, would aid in explaining its relatively firm retention in the complex. Secondly, the need for a more or less rigid orientation of the aminoacyl-tRNA on the ribosome also requires the assumption of supplementary binding of tRNA by a region differing from the anticodon. In the third place, a certain affinity of tRNA for the ribosome itself can be detected in experiments; for example, at pH of about 5 to 6 appreciable binding of aminoacyl-tRNA to the 30 S subparticle of the ribosome is observed in the absence of the template [PESTKA, NIRENBERG, 1966 (1)].

d) Ribosomal Site Interacting with the Codon-Anticodon Pair and tRNA

As has already been mentioned, the ribosome possesses only one functional site binding the template polynucleotide, and the indicated region is localized on the 30 S subparticle. In full agreement with this, it was found that isolated 30 S subparticles, like whole 70 S ribosomes, can form a ternary complex with the template polynucleotide and aminoacyl-tRNA [MATTHAEI et al., 1964; SUZUKA et al., 1965; KAJI et al., 1966 (2); PESTKA, NIRENBERG, 1966 (1, 2)]. The complex of 30 S–template–aminoacyl-tRNA is formed most completely at a Mg^{++} ion concentration of about 0.02 M, although at 0.01 M Mg^{++}, complex formation is also rather substantial. The binding of aminoacyl-tRNA into a complex with the 30 S subparticle and the template polynucleotide is strictly specific: the selection of a given aminoacyl-tRNA is entirely determined by the template polynucleotide or oligonucleotide in exact correspondence to the code. Both aminoacyl-tRNA and the corresponding free (deacylated) tRNA are bound into the complex. Increased temperatures promote the formation of the complex 30 S–template–aminoacyl-tRNA: at 37° the rate of binding of aminoacyl-tRNA is rather great, at 24° it is twice less, at 15° it is many times lower, while at temperatures from 0 to 5°, binding is already very difficult to observe. K^+ and NH_4^+ ions in a concentration of 0.1 M *stimulate* the binding of aminoacyl-tRNA in the system 30 S subparticle + template, while Na^+ and Li^+ either have no appreciable influence or (especially Li^+) reduce the binding. Without the template polynucleotide, the 30 S subparticle has practically no ability to bind aminoacyl-tRNA under normal conditions (0.01 to 0.02 M Mg^{++}, pH 7 to 9). Thus, in all the important features, the conditions of formation of the 30 S–template–aminoacyl-tRNA complex are identical with the conditions of formation of the usual ternary complex of ribosome–template–aminoacyl-tRNA (see above, Section 2, a). This cannot but indicate that it is the 30 S subparticle that bears the main responsibility for the specific binding of aminoacyl-tRNA on the ribosome in the presence of the template. Consequently, it is precisely on the 30 S subparticle that the site, binding at least the codon-anticodon pair, and possibly the remainder of the tRNA molecule as well, must be localized.

It should be mentioned that the addition of the 50 S subparticles to the system of 30 S + template stimulates the specific binding of aminoacyl-tRNA to a greater or lesser degree [SUZUKA et al., 1966; PESTKA, NIRENBERG, 1966 (1)]. This can evidently occur for two reasons: 1. on the 50 S subparticle there may be *supplementary forces* of retention of tRNA on the ribosome; 2. the association of the 50 S subparticle with the 30 S subparticle into the complete ribosome can stabilize the *appropriate* conformation of the 30 S subparticle and its tRNA-binding site.

In any case it can be concluded that the *aminoacyl-tRNA-binding site* of the ribosome, or the site of *specific* binding of tRNA is situated on the 30 S subparticle (see Fig. 18). The center retaining the codon-anticodon pair (it is probably the same as the mRNA-binding center of the ribosome) may be considered as a part of the indicated aminoacyl-tRNA-binding site.

It cannot yet be stated with assurance which of the two chemical components of the 30 S subparticle, the ribosomal RNA or the ribosomal protein plays the main role in the binding of the codon-anticodon pair and the remainder of the tRNA molecule, i.e., in the formation of centers of the aminoacyl-tRNA-binding site. From general considerations, we might rather give preference to the *protein component*, as responsible for the binding of tRNA. If the center binding the codon-anticodon pair is identical with the mRNA-binding center noted above (Section 1, c), then here are also certain experimental, albeit indirect, indications of the role of the protein component (see Section 1, c). Finally, a direct experimental indication of the role of the amino group of the ribosomal protein in the binding of tRNA has also been obtained [MOORE, 1966 (2)].

e) Summary

In the presence of template polynucleotide and aminoacyl-tRNA in the system the ribosome forms a stabilized ternary ribosome–template–aminoacyl-tRNA com plex. Only aminoacyl-tRNA, the specificity of which corresponds to the codons o the template polynucleotide, is strictly selectively bound in the complex (the so-called "specific binding" of aminoacyl-tRNA). The presence or absence of the aminoacyl residue itself, however, is not of decisive significance. The single-stranded region of the template polynucleotide participates in the association; the phosphorylated 5′-end region in this case evidently has an advantage over the internal regions of the polynucleotide in the formation of the complex. The complex formed is cooperative in the sense that the binding of aminoacyl-tRNA stabilizes the retention of the polynucleotide, while the retention of the polynucleotide, in turn, ensures binding of aminoacyl-tRNA. The formation of the ternary complex requires a Mg^{++} concentration in the medium of about 0.005 to 0.015 M. K^+ and NH_4^+ ions stimulate association, whereas Na^+ and Li^+ ions are either inactive or suppress it. At the lower Mg^{++} concentration the process of aminoacyl-tRNA binding with the ribosome is promoted by the protein transfer factors (T) and GTP (without its cleavage).

The basis of the ternary association is evidently complementary interaction of the codon of the template polynucleotide with the anticodon of tRNA and firm retention of this codon-anticodon pair on the ribosome. The indicated retention of the codon-anticodon pair and formation of the ternary complex are provided for chiefly by the 30 S subparticle of the ribosome. The center of the 30 S subparticle, binding the codon-anticodon pair, may be identical with the mRNA-binding center of the 30 S subparticle in the formation of a binary ribosome–template complex. The possibility remains that the 30 S subparticle may contain additional centers of tRNA binding, which in conjunction with the codon-anticodon-binding center form the *aminoacyl-*

tRNA-binding site. It may be assumed that the main responsibility for the formation of this binding site on the 30 S subparticle is borne by the ribosomal protein.

3. Association of the Ribosome with tRNA in the Absence of a Template

It has been shown that the ribosomes of *E. coli* can bind tRNA or aminoacyl-tRNA in the absence of template polynucleotide as well (CANNON et al., 1963; CANNON, 1967). In contrast to selective binding of specific tRNA in the presence of the template, this binding is sometimes denoted as "nonspecific binding" of tRNA by the ribosome.

Peptidyl-tRNA's, added to ribosomes in the absence of template polynucleotide, exhibit far more active "nonspecific" binding than free tRNA's or aminoacyl-tRNA's [RYCHLIK, 1966; PESTKA, NIRENBERG, 1966 (2)].

a) Conditions of Association and Stability of the Complex

The binding of tRNA or aminoacyl-tRNA by purified ribosomes without a template requires only a Mg^{++} concentration of about 0.01 M. Various tRNA's and aminoacyl-tRNA's are evidently bound to an equal degree, i.e., the ribosome does not prefer any of them. In the presence of 10^{-4} M Mg^{++}, the ribosome–tRNA complex dissociates entirely. A characteristic feature of the complex under consideration is the fact that its formation is *suppressed* by K^+ ions, even when a high concentration of Mg^{++} ions is maintained. The complex is readily formed in the cold (0 – 4 °C) as well. Neither energy sources nor protein factors are required.

Bound deacylated tRNA or aminoacyl-tRNA readily exchanges with exogenous tRNA's. The excess deacylated tRNA readily displaces aminoacyl-tRNA from the ribosome, and *vice versa*. Consequently, a dynamic equilibrium exists:

$$\text{ribosome} + \text{tRNA} \rightleftharpoons \text{ribosome–tRNA}.$$

The very labile binding and easy exchangeability of bound tRNA are evidently the important distinguishing features of this complex in comparison with the so-called "specific" binding of tRNA in the presence of the template (see above, Section 2).

The binding of peptidyl-tRNA to the ribosome is evidently far stronger (RYCHLIK, 1966). It also requires a Mg^{++} concentration of about 0.01 M, but the monovalent cations K^+ and NH_4^+ in a concentration of 0.1 M do not suppress the formation of the complex in this case. Binding is only partially suppressed by Na^+ and Li^+ ions. The complex is well formed at 0°, and increasing the temperature to 25° only slightly accelerates complex formation. Neither energy sources nor protein factors are required here either. It may be thought that the nonspecific binding of the peptidyl-tRNA to the ribosome is analogous in principle to the nonspecific binding of the free tRNA's and aminoacyl-tRNA's, with the exception that there is a substantial difference in the strength of the binding.

b) Possible Regions of tRNA Interacting with the Ribosome

The available experimental data (CANNON et al., 1963) permit us to speak with certainty only of the role of the universal 3′-terminal CCA sequence of the tRNA molecules (Fig. 17) in the "nonspecific binding" to the ribosome. A disruption of this terminal sequences leads to disappearance of the affinity of tRNA for the ribosome. Therefore the terminal CCA region may be at least one of the main points of attachment of tRNA or aminoacyl-tRNA to the ribosome in the case of nonspecific binding.

A possible role of other regions of the tRNA molecule in nonspecific binding with the ribosome also cannot be excluded. Since this binding does not depend upon the specificity of tRNA, such regions should most likely be sought among universal nucleotide combinations, common to all tRNA chains. For example, a universal sequence for all tRNA's is found in the region of the 50 to 60th nucleotide of the chain: GTΨC (Fig. 17). A definite similarity among different tRNA's is also observed in the region around the 30th nucleotide of the chain, where most of the dihydrouridine residues are concentrated (Fig. 17). It would be interesting to determine whether the "ribothymidyl-pseudouridyl" and "dihydrouridyl" regions of tRNA have some function in the "nonspecific" binding of tRNA or aminoacyl-tRNA to the ribosome.

As has already been indicated, both free (deacylated) and aminoacylated tRNA's can be bound to the ribosome, i.e., the presence of an aminoacyl residue on the 3′-end the chain does not radically change the affinity of the molecule for the ribosome. However, if instead of an aminoacyl residue, a peptidyl, even a very short peptidyl, with only two or three amino acid residues, is present on the 3′-end of the tRNA chain, then the affinity of the entire molecule for the ribosome is sharply increased (Cannon, 1967). The ribosome–peptidyl-tRNA complex proves to be stable not only at 10^{-2} M Mg^{++}, but even at 10^{-4} M Mg^{++}, both in the absence and in the presence of K^{+} ions; replacement of peptidyl-tRNA's by exogenous tRNA's used in a large excess, is not observed, i.e., the binding constant evidently is very high. It may be that in this case the first (from the 3′-end of the tRNA chain) peptide group represents a point of strong supplementary binding of the peptidyl-tRNA molecule to the ribosome. However, another possibility also exists: the peptidyl residue may induce a transition of tRNA to a new conformational state, characterized by an especially great affinity for the ribosome.

c) Ribosomal Site Interacting with tRNA

The available experimental data (Cannon et al., 1963; Cannon, 1967) show that evidently there is only *one site* on the ribosome for the "nonspecific" reversible binding of tRNA or aminoacyl-tRNA being analyzed. This same site is responsible for the firm binding of peptidyl-tRNA (Cannon, 1967).

In experiments with isolated subparticles, it is shown that only the 50 S subparticle, and not the 30 S subparticle, binds tRNA or aminoacyl-tRNA in the absence of the template polynucleotide. In exactly the same way, if the ribosome contains peptidyl-tRNA, it remains bound to the 50 S subparticle after dissociation. Consequently, the site binding tRNA in the absence of the template is localized *on the 50 S subparticle* of the ribosome, in all probability, on its surface adjoining the 30 S subparticle (Fig. 18). We shall henceforth denote this site as the *peptidyl-tRNA-binding site* of the ribosome.

It was recently shown that the indicated binding of tRNA on the 50 S subparticle depends upon a definite small fraction of the structural ribosomal protein of this subparticle (NOMURA, TRAUB, 1966; RASKAS, STAEHELIN, 1967). When this fraction is removed in the process of disassembly of the subparticle, the ability to bind tRNA is lost; if, however, this group of proteins is combined with the subparticle, then the ability to bind tRNA is restored, even in the absence of a number of other structural proteins. Thus, it might be thought that the protein component bears the principal responsibility for the formation of the tRNA-binding site on the 50 S subparticle.

Fig. 18. Schematic picture of the ribosome (the component subparticles are somewhat spread apart to show the contacting surfaces), associated with mRNA (on the 30 S subparticle) and with peptidyl-tRNA (on the 50 S subparticle); aminoacyl-tRNA is entering the 30 S subparticle

d) Summary

Purified ribosomes can bind deacylated tRNA or aminoacyl-tRNA in the absence of all other high-molecular weight components of the protein-synthesizing system, and also in the absence of template polynucleotides. This binding requires only a definite concentration of divalent cations, Mg^{++}, in the medium — about 0.01 M. K^+ ions suppress such binding. The specificity of tRNA and the presence or absence of the aminoacyl residue are not of decisive importance. Bound tRNA or aminoacyl-tRNA exists in a dynamic equilibrium with exogenous tRNA's and is readily exchanged with them. Peptidyl-tRNA is far more firmly retained in the association with the ribosome in the absence of a template.

There is only one site on the ribosome that binds tRNA in the absence of the template. This site is situated on the 50 S subparticle and can be denoted as the peptidyl-tRNA-binding site. It is most likely formed by a definite group of ribosomal proteins. Evidently the tRNA residue of the peptidyl-tRNA molecule, firmly retained by the working ribosome, is situated precisely in this site.

4. Specific Stimulation of the Binding of Peptidyl-tRNA to the Ribosome by the Template

It has been shown that although peptidyl-tRNA (in particular polylysyl-tRNA) binds to the ribosome in the absence of the template, the addition of the template polynucleotide corresponding to it (polyA) appreciably stimulates the binding [RYCHLIK, 1966; PESTKA, NIRENBERG, 1966 (2)]. It might be thought that the entering of the peptidyl-tRNA on the peptidyl-tRNA-binding site of the 50 S subparticle is considerably facilitated if it passes *through preliminary specific binding with the template polynucleotide and aminoacyl-tRNA-binding site on the 30 S subparticle.* In any case this looks like one of the reasonable explanations for the data cited above on stimulation. The possibility of a specific binding of peptidyl-tRNA with the 30 S subparticle in the presence of a corresponding template was experimentally demonstrated [PESTKA, NIRENBERG, 1966 (2)]. Moreover, it was shown that the template polynucleotide not corresponding to the investigated peptidyl-tRNA in code specificity, practically completely suppresses its nonspecific binding with the complete (70 S) ribosome although it does not influence its binding with the isolated 50 S subparticle [PESTKA, NIRENBERG, 1966 (2)].

The latter indicates that the presence of the template in the complete ribosome must also fully suppress the *nonspecific* binding of aminoacyl-tRNA with the peptidyl-tRNA-binding site of the 50 S subparticle. Even if the aminoacyl-tRNA in the presence of the template can reach the peptidyl-tRNA-binding site of the complete (70 S) ribosome it does so only through preliminary specific binding with the template on the aminoacyl-tRNA-binding site of the 30 S subparticle. This means that any binding at both tRNA-binding sites of the ribosome *in the presence of a template can be only specific.* There are indirect experimental indications in support of the possibility of such specific binding of two molecules of aminoacyl-tRNA on the ribosome: the amount of specifically bound aminoacyl-tRNA is the same when tri-, tetra-, and pentanucleotides are used as the template and increases appreciably in the case of hexanucleotides (ROTTMAN, NIRENBERG, 1966). However, it should be emphasized that such binding of aminoacyl-tRNA was always studied at Mg^{++} concentrations evidently substantially exceeding the physiological, for example, in the presence of 0.03 M Mg^{++}. Under physiological concentrations of Mg^{++}, from 0.005 M to 0.01 M, the affinity of the aminoacyl-tRNA for the peptidyl-tRNA-binding site of the 50 S subparticle is so insufficient, that binding of usual aminoacyl-tRNA's with the peptidyl-tRNA-binding site is probably not observed in the presence of a template. However, the blocking of the NH_2-group of the aminoacyl-tRNA leads, evidently, to the considerable increase of affinity of the aminoacyl-tRNA for the peptidyl-tRNA-binding site of the 50 S subparticle, so that in the presence of the corresponding template, the N-blocked aminoacyl-tRNA's, just as the peptidyl-tRNA, specifically binds to it, probably through the preliminary specific binding with the aminoacyl-tRNA-binding site of the 30 S subparticle. The initiator tRNA, formylmethionyl-tRNA, belongs to such N-blocked aminoacyl-tRNA's (see Section III, 1).

III. Stages of Translation

In the complete protein-synthesizing system, i.e., in the presence of all the necessary components enumerated above (Section I), the ribosome performs its function of *translation*: synthesis of the polypeptide chain of a protein takes place, and the amino acid sequence of the chain synthesized is unambiguously determined by the nucleotide sequence of mRNA. The totality of processes carried out by the ribosome and leading ultimately to the formation of a completed polypeptide chain of a protein, can be divided into three successive stages: 1. *initial association* of the ribosome with the template and aminoacyl-tRNA and *initiation of translation*; 2. translation itself, or *polymerization of amino acid residues into a polypeptide chain*; 3. cessation of polymerization of the amino acid residues and release of the polypeptide chain of the protein from the ribosome, i.e., *termination of translation.*

During the course of translation, according to modern concepts, the ribosome does not accommodate the entire template polynucleotide, but successively draws it from one end to the other, at each given moment being bound only to a definite, very limited section of a template [WATSON, 1963; GILBERT, 1963 (1); WARNER et al., 1963; RICH et al., 1963; GIERER, 1963; NOLL et al., 1963; SPIRIN, 1964]. Simultaneously with the drawing of the polynucleotide along the ribosome, the polypeptide chain is synthesized. Thus, the ribosome reads out the information encoded in the form of the linear sequence of nucleotides of the mRNA chain, continuously *being displaced relative to this chain.*

It is known that a polynucleotide is a polar polymer: each internucleotide bond is formed in such a way that the phosphate residue connects the 3′-position of one nucleoside to the 5′-position of the neighboring nucleoside, and this direction of the internucleotide bonds is the same along the entire chain of the polynucleotide (Fig. 4). The nucleotide sequence in the chain is customarily read in the direction of the $C_{3'}$-P-$C_{5'}$ linkage. Then the 5′-position in the first nucleotide residue of the chain will be unoccupied in the internucleotide linkage, while in the last nucleotide residue of the chain, the 3′-position will be free. Correspondingly, the first nucleotide residue of the chain is customarily denoted as the 5′-end of the polynucleotide, and the final nucleotide residue as the 3′-end. It has now been shown that during the process of translation, the *template polynucleotide is read by the ribosome always in the direction of the $C_{3'}$-P-$C_{5'}$ internucleotide linkage, i.e., from the 5′-end to the 3′-end* (THACH et al., 1965; SALAS et al., 1965; TERZAGHI et al., 1965, 1966; SMITH et al., 1966; LAMFROM et al., 1966).

The two premises indicated — the dynamic movement of the chain of the template polynucleotide relative to the ribosome and polarity of the movement, from the 5′-end to the 3′-end, — are fundamental in a consideration of all stages of translation.

1. Formation of the Initial Complex Ribosome — Template — (Aminoacyl-tRNA)$_2$ and Initiation of Translation

The initiation of translation itself means the synthesis of the first peptide bond in the initial complex, in which a ribosome, template polynucleotide, and aminoacyl-tRNA molecules must participate. Obviously the presence of more than one aminoacyl in the complex is needed for the synthesis of the peptide bond. Evidently the ribosome principally ensures initiation of translation by having two different tRNA-binding sites (see Section II, 2, d and 3, c).

However, as was indicated above (Section II, 4), the presence of the template polynucleotide ensures the specific binding of the aminoacyl-tRNA to the aminoacyl-tRNA binding site of the 30 S subparticle, but under conditions close to physiological the second tRNA-binding site, the peptidyl-tRNA-binding site of the 50 S subparticle, has no sufficient affinity to bind the usual aminoacyl-tRNA. Thus, despite the presence of two tRNA-binding sites on the ribosome, the free binding of the second molecule of aminoacyl-tRNA with the ribosome appears to be practically impossible. Hence, the basic question of the initiation of translation arises — *how the simultaneous involvement of two aminoacyl-tRNA's into a complex with the ribosome and the template polynucleotide is ensured.*

It was shown that under physiological conditions the only manner by which this question is settled, is the use of the N-blocked aminoacyl-tRNA which has an increased affinity for the peptidyl-tRNA-binding site of the ribosome. Consequently, it is the N-blocked aminoacyl-tRNA that serves as an *initiator tRNA* in the protein synthesis. In bacterial systems the *N-formylmethionyl-tRNA* serves as such a N-blocked aminoacyl-tRNA which ensures the initiation of translation.

a) The Initiator Formylmethionyl-tRNA

The N-formylmethionyl-tRNA was discovered in *E. coli* by MARCKER and SANGER in 1964 (MARCKER, SANGER, 1964).

It has been shown that only a strictly specific $tRNA_{f\text{-}met}$ participates in the formation of formylmethionyl-tRNA. It first accepts methionine in the usual way, forming methyionyl-$tRNA_{f\text{-}met}$, but then a special enzyme system of *E. coli* formylates the NH_2 group of the methionine residue of this methionyl-$tRNA_{f\text{-}met}$ (MARCKER, 1965). Free methionine or methionyladenylate is not formylated. Moreover, if N-formylmethionine is produced artificially, it is not accepted by $tRNA_{f\text{-}met}$.

In addition to $tRNA_{f\text{-}met}$, on which formylation of methionine occurs after it is accepted, *E. coli* also contains the usual $tRNA_{met}$, which also accepts methionine, but without its subsequent formylation [CLARK, MARCKER, 1966 (1); KELLOGG et al., 1966]. Consequently, the formylating enzyme system specifically recognizes not simply the methionyl residue, but rather the corresponding $tRNA_{f\text{-}met}$ to which it is bound.

The role of the formylmethionyl-tRNA as an initiator tRNA was at first substantiated in experiments, where it was shown that in the protein-synthesizing system of *E. coli* its formylmethionyl residue proved to be permanently in the N-end position of the synthesized polypeptide chains [ADAMS, CAPECCHI, 1966; CAPECCHI, 1966 (1, 2); WEBSTER et al., 1966]. Since it is known that the polypeptide chain of the protein during the process of translation grows from the N-end to the C-end, the N-end amino acid of the chain must be the first from which the growth of the chain in the ribosome begins. Hence, the N-formylmethionyl-tRNA is the first aminoacyl-tRNA which determines the incorporation of the N-end residue and thereby commences the process of translation.

Evidently, since the NH_2-group in formylmethionyl-tRNA is masked, this aminoacyl-tRNA can serve as nothing else but for the positioning of the amino acid on the very N-end position of the polypeptide chain, i.e., for the initiation of translation. For the transfer of methionine residue to any other position of the polypeptide chain except the N-end position, another methionine tRNA, a usual methionyl-$tRNA_{met}$ is employed. Thus, the masking of the NH_2-group in the formylmethionyl-tRNA distinguishes it from other aminoacyl-tRNA's and excludes the possibility for it to transfer aminoacyl inside the growing peptide chains, thus permitting it only to begin new chains from the N-end position.

It is apparent that the initiator function of the formylmethionyl-tRNA is finally provided for by its *particular*, *increased affinity for the peptidyl-tRNA-binding site of the ribosome* (BRETSCHER, MARCKER, 1966). In the presence of a template polynucleotide all the usual aminoacyl-tRNA's under physiological conditions can, in all probability, be bound and retained only in the aminoacyl-tRNA-binding site (on the 30 S subparticle) but not in the peptidyl-tRNA-binding site (on the 50 S subparticle). That is why, only the presence of the formylmethionyl-tRNA in the system leads to the fact that two aminoacyl-tRNA's, one of which is the formylmethionyl-tRNA, may simultaneously be found within the ribosome.

However, the detailed mechanism of the initiation of translation is essentially more complicated and specific than is the simple and independent entering of the initiator tRNA into the peptidyl-tRNA-binding site and the entering of another aminoacyl-tRNA into the aminoacyl-tRNA-binding site. Quite a number of other specific factors participate in this initiation mechanism with the result that the process of initiation passes through several successive stages. As a result, the natural initiation is a very specific and well controlled process.

b) Special Factors and Conditions Ensuring the Formation of the Initial Complex and Initiation of Translation

Initiation Codons. For the effective binding of the initiator formylmethionyl-tRNA into the initial complex, the template polynucleotide of the complex must have a special nucleotide combination. Nucleotide combinations coding formylmethionyl-tRNA, i.e., ensuring its binding into the initial complex, are, in the main, the triplets AUG, and GUG [CLARK, MARCKER, 1966 (1); KELLOGG et al., 1966; SUNDARARAJAN, THACH, 1966; GHOSH et al., 1967].

The triplets AUG and GUG code formylmethionine only when they are the *initial* triplets in the reading out of the template polynucleotide. If, however, these nucleotide triplets are encountered in the chain of a template polynucleotide that has already begun to be read out, when they are the usual "internal" triplets, each of them codes its own amino acid: AUG — methionine, GUG — valine (see Fig. 5). Correspondingly, the initial AUG triplet of the polynucleotide chain associated with a ribosome leads to selective binding of formylmethionyl-$tRNA_{f\text{-}met}$, while the "internal" AUG triplet reacts only with methionyl-$tRNA_{met}$ during the course of translation. Therefore, if the regular synthetic polymer $poly(AUG)_n$ is used as the template polynucleotide in a cell-free protein-synthesizing system, then the polypeptide N-formylmet-met-met-met.... is formed (GHOSH et al., 1967). With the regular polymer $poly(UG)_n$ as the template the polypeptide N-formylmet-cys-val-cys-val-.... is formed (the first GUG codes formylmet, UGU codes cys, and the internal GUG codes val — see Fig. 5) (GHOSH et al., 1967).

Involvement of the 5′-end of the Template Polynucleotide in the Formation of the Initial Complex. In a consideration of the formation of the initial complex, the decisive factor is that the ribosome must in all probability exhibit special affinity for the 5′-end of the template polynucleotide. In any case, in the initial association of the ribosome with mRNA *in vivo*, as well as with the same natural mRNA's in cell-free systems *in vitro*, the ribosomes evidently do not attach to the internal sections of the mRNA chain.

This conclusion follows from various groups of data. In the first place, in experiments *in vitro* with polyU, the predominant association of the ribosome with the end of the template polynucleotide chain was demonstrated by electron microscopy (MATTHAEI et al., 1964). In the second place, all the experience in the use of viral RNA's as natural mRNA's in protein-synthesizing cell-free systems of *E. coli* (see, for example, OHTAKA, SPIEGELMAN, 1963; CLARK et al., 1965) indicates that only complete polypeptide chains of the viral proteins are formed, i.e., the mRNA's are read out from the beginning, and the ribosome never begins translation from the internal sections of the mRNA cistrons. In the third place, from data on the kinetics of the appearance of the corresponding proteins in experiments *in vivo*, it can be concluded that polycistronic mRNA's are read by the ribosomes only from the "operator" end (for example, KEPES, 1967). Finally, the most indicative are the genetic data: a "nonsense" mutation in one of the first cistrons of polycistronic mRNA, interrupting the translation of this cistron, also leads to suppression of translation of the other cistrons situated in the direction of the 3′-end (in the direction of the reading) from the mutant cistron; this polar effect of "nonsense" mutation agrees but with the assumption that the ribosome can combine only with the 5′-end of the polycistronic mRNA and reads it strictly in sequence, cistron by cistron, in the direction from the 5′-end to the 3′-end of the chain (YANOFSKY, ITO, 1966; MARTIN et al., 1966).

Consequently, the initiator codon of the template polynucleotide, forming an initial complex with the ribosome and formylmethionyl-tRNA, evidently should always be only that AUG (or GUG) combination which is situated at the 5′-end of the polynucleotide.

However, for the formation of the initial complex, the AUG (or GUG) combination does not at all necessarily have to be at the *very* 5′-end of the template polynucleotide. For example, if the polymer AAUG$(U)_{25}$ or AAAUG$(U)_{25}$ is taken as the template in the protein-synthesizing system *in vitro*, polyphenylalanine polypeptides (UUU codes phenylalanine — Fig. 5) with N-formylmethionine at the N-end will be synthesized in exactly the same way as with the polymer AUG$(U)_{25}$ (THACH et al., 1966). This means that the initiation codon may be somewhat apart from the 5′-end; then during association the ribosome may somehow miss one to two nucleotides preceding the initiation codon and bind just the same into the initial complex with the first AUG combination. Thus, the reading of the template polynucleotide can be accomplished not simply from the very 5′-end, but in such a way that the initiation triplet is read first, whether it is at the very 5′-end or at some distance from it; then the following nucleotide sequence is always read by triplets, *in phase* from the initiator triplet.

And yet, even the indicated removal of the initiator triplet by only one or two nucleotides from the 5′-end of the template polynucleotide substantially reduces its activity in the formation of the initial complex (approximately by half) in comparison with the 5′-terminal AUG triplet (THACH et al., 1966). Consequently, the 5′-end itself is an extremely important factor ensuring effective formation of the initial complex. The initial association of the ribosome with the 5′-end of mRNA is specifically stimulated by a special protein factor — the "initiation factor F 3" (see Section I, 6). The possibility remains that in natural mRNA's the initiation codon always occupies the very 5′-terminal position.

Cations, Protein Factors and GTP. In addition to the presence of formylmethionyl-tRNA in the system and the presence of an initiation codon at the 5′-end of the template polynucleotide, the formation of the specific initial complex and normal initiation of translation require at least three more conditions.

In the first place, a definite concentration of Mg^{++} ions must be maintained in the medium within rather narrow limits, from 0.005 M to 0.009 M, *no higher* (NAKAMOTO, KOLAKOFSKY, 1966; KOLAKOFSKY, NAKAMOTO, 1966; SUNDARARAJAN, THACH, 1966). At higher Mg^{++} concentrations, the association of the ribosome with the template and the usual tRNA's becomes stronger, and then the ribosomes begin to form the initial complex less specifically, independent of whether there is an initiation codon or not, and whether formylmethionyl-tRNA is present or not. In other words, at high concentrations of Mg^{++} there exists a great probability of binding the usual aminoacyl-tRNA with the peptidyl-tRNA-binding site of the ribosome. (Precisely such a situation occurs when synthetic template polynucleotides containing no initiator triplets are used in a cell-free system; in this case the beginning of synthesis of the polypeptide requires increased concentrations of Mg^{++}, from 0.01 M to 0.02 M).

In the second place the presence of protein "initiation factors" F1 and F2 is important for the formation of a specific initial complex with the participation of the initiating triplet and formylmethionyl-tRNA, and for subsequent initiation of translation (see above, Section I, 6). In all probability their role consists, at least partially, of ensuring the appropriate binding of the formylmethionyl-tRNA with the ribosome in the presence of the initiating triplet (AUG or GUG).

In the third place, in addition to the initiating codon and the "initiation factors" F1 and F2, GTP is also required for the binding of the formylmethionyl-tRNA with the ribosome (HERSHEY and THACH, 1967; LUCAS-LENARD and LIPMANN, 1967; THACH et al., 1967; OCHOA et al., 1967; ALLENDE and WEISSBACH, 1967; ANDERSON et al., 1967; LEDER and NAU, 1967; OHTA et al., 1967; HILLE et al., 1967).

c) Formation of the Initial Complex with Participation of the Initiator tRNA and the Initiating Codon

Although, as has already been mentioned, the initiator formylmethionyl-tRNA possesses, *in contrast to all the usual aminoacyl-tRNA*, a substantial affinity for the peptidyl-tRNA-binding site on the ribosome (BRETSCHER, MARCKER, 1966; SUNDARARAJAN, THACH, 1966; NAKAMOTO, KOLAKOFSKY, 1966), the presence of the initiating codon bound to the ribosome is firstly required for its effective entering into the ribosome (BRETSCHER, MARCKER, 1966; ZAMIR et al., 1966).

The stimulation of binding of formylmethionyl-tRNA with the peptidyl-tRNA-binding site of the ribosome in the presence of the initiating codon may be explained by the necessity of its passing through preliminary specific binding at the aminoacyl-tRNA-binding site on the 30 S subparticle.

$$\begin{array}{c} 50\,S \\ | \\ 30\,S \end{array} + \text{f-met-tRNA} + \text{AUG} \rightleftharpoons$$

$$\rightleftharpoons \left(\begin{array}{l} 50\,S \\ | \\ 30\,S - \text{AUG} \end{array} + \text{f-met-tRNA} \right) \rightleftharpoons$$

$$\rightleftharpoons \left(\begin{array}{l} 50\,S \\ | \quad \diagup \text{f-met-tRNA} \\ 30\,S \quad\quad | \\ \quad\quad \diagdown \text{AUG} \end{array} \right) \rightarrow$$

$$\rightarrow \begin{array}{lcl} 50\,S & — & \text{f-met-tRNA} \\ | & & | \\ 30\,S & & \text{AUG} \end{array}$$

This mechanism was proposed above to explain the stimulation of the binding of peptidyl-tRNA by the template (Section II, 4).

The fact that the isolated 30 S subparticles actively bind the formylmethionyl-tRNA in the presence of the initiating codon, protein "initiation factors" and GTP, supports this suggestion (NOMURA, LOWRY, 1967; HILLE et al., 1967). One can assume that in the abovementioned case the formylmethionyl-tRNA is bound on the aminoacyl-tRNA-binding site. The GTP is not cleaved in this case as it is during the translation, but serves most likely in the role of an allosteric effector for the 30 S subparticle (OCHOA et al., 1967; HILLE et al., 1967; OHTA et al., 1967). The role of the protein "initiation factors" F1 and F2 may consist of ensuring the supplementary affinity of the N-blocked aminoacyl-tRNA (formylmethionyl-tRNA) for the aminoacyl-tRNA-binding site of the 30 S subparticle. Consequently the primary association of formylmethionyl-tRNA occurs, most probably, with the 30 S subparticle and not with the 50 S subparticle (NOMURA, LOWRY, 1967; HILLE et al., 1967).

Moreover, the evidence obtained indicates that the primary association of the formylmethionyl-tRNA in the presence of the template with the initiating codon occurs only with the separate 30 S subparticles but not with the 30 S subparticles within the complete 70 S ribosomes (NOMURA, LOWRY, 1967); it is probably only after this primary initial association of the 30 S subparticles with formylmethionyl-tRNA that the formed ternary complex, 30 S subparticle–template–formylmethionyl-tRNA, associates with the 50 S subparticle into a complete ribosome (NOMURA, LOWRY, 1967; NOMURA et al., 1967; SCHLESSINGER et al., 1967).

Nonetheless in the initial complex ribosome–template–(aminoacyl-tRNA)$_2$ the formylmethionyl-tRNA is ultimately found in the peptidyl-tRNA binding site (BRETSCHER, MARCKER, 1966; ZAMIR et al., 1966; OHTA et al., 1967; HILLE at al., 1967), i.e., in all probability on the 50S subparticle. Hence, it may be assumed that the initial association with the aminoacyl-tRNA binding site on the 30S subparticle is followed by the transition (translocation) of the formylmethionyl-tRNA to the peptidyl-tRNA binding site of the 50 S subparticle.

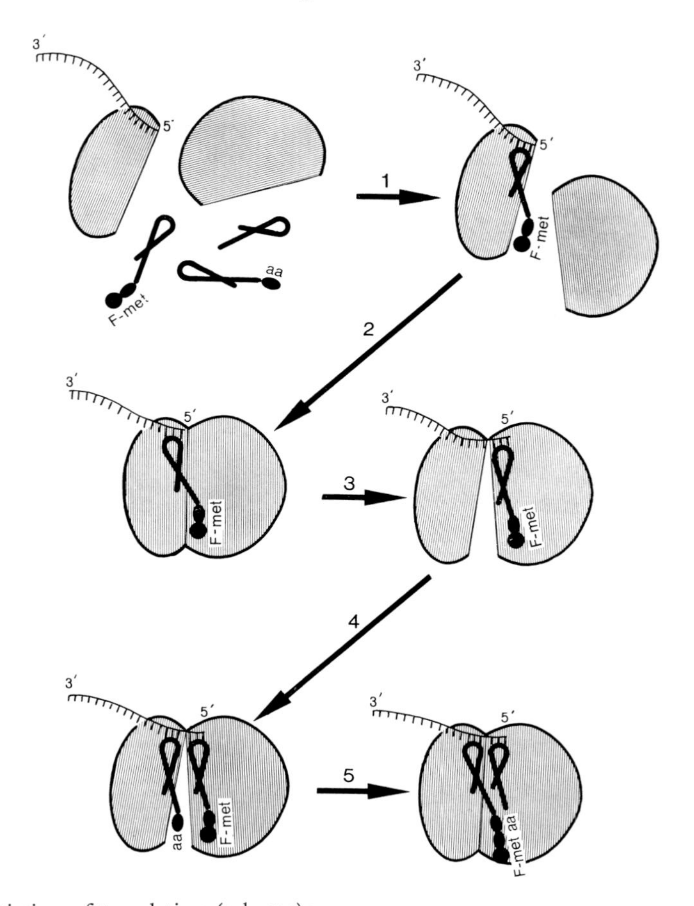

Fig. 19. Initiation of translation (scheme):

1 Binding of the formylmethionyl-tRNA with an initial codon of the template polynucleotide on the 30 S subparticle;
2 Association of subparticles into a complete ribosome;
3 Translocation of the formylmethionyl-tRNA to the peptidyl-tRNA-binding site of the 50 S subparticle;
4 Binding of the second aminoacyl-tRNA to the aminoacyl-tRNA-binding site of the 30 S subparticle;
5 Formation of the first peptide bond: transfer of the formylmethionyl residue from the $tRNA_{f\text{-}meth}$ to the amino group of the aminoacyl-tRNA.

As a whole the following scheme of the sequence of events during the formation of the initial complex can be proposed. *Step 1*: a free 30 S subparticle associates loosely with the 5′-end of the template polynucleotide (Eisenstadt, Brawerman, 1967; Gros, 1967). A special "initiator factor" of a protein nature (Revel, Gros, 1967; Gros, 1967) stimulates this association. *Step 2* (Fig. 19-1): if there is a AUG (or GUG) codon at the 5′-end of the template, then the formylmethionyl-tRNA associates with this codon on the 30 S subparticle, settling on the aminoacyl-tRNA-binding site and forming a stable ternary complex of the 30 S–AUG–formylmethionyl-tRNA. Two protein "initiation factors", F1 and F2, and also GTP promote the binding of this unusual tRNA with the aminoacyl-tRNA-binding site. This binding sets the phase for the further reading of the template polynucleotide in that it positions the AUG (or GUG) codon exactly in the mRNA-binding center of the aminoacyl-tRNA-binding site, no matter whether this codon is at the very 5′-end of the template or one or two nucleotides away. *Step 3* (Fig. 19-2): the 50 S subparticle associates with the 30 S subparticle which retains the formylmethionyl-tRNA. This association can be caused by the affinity of the 50 S subparticle to the formylmethionyl-tRNA imitating the peptidyl-tRNA; this affinity makes an additional contribution to the stability of the 50 S-to-30 S complex ensuring the shift of the equilibrium of the 50 S + 30 S $\rightleftarrows$ 70 S reaction towards the 70 S. *Step 4* (Fig. 19-3): the formylmethionyl-tRNA, together with the AUG (or GUG) codon retained by it, passes from the aminoacyl-tRNA-binding site of the 30 S subparticle onto the peptidyl-tRNA-binding site of the 50 S subparticle in the complete ribosome. Thus, this means the *translocation* of the formylmethionyl-tRNA and a shift of the template polynucleotide by one triplet. Cleavage of the GTP molecule is probably necessary for this translocation. (This explains the observation that the GTP cleavage is required for the formation of the first peptide bond and occurs after the binding of the formylmethionyl-tRNA, but before the formation of the peptide bond — see Hershey, Thach, 1967; Thach et al., 1967; Ohta et al., 1967). As a result of the tRNA translocation and the corresponding drawing over of the template, the aminoacyl-tRNA-binding site becomes vacant and a triplet following the initiating AUG (or GUG) triplet is positioned on it. *Step 5* (Fig. 19-4): the aminoacyl-tRNA-binding site of 30 S subparticles, with the positioned triplet adjacent to the initiating codon, selectively binds the aminoacyl-tRNA corresponding to this triplet. As a consequence the initial complex ribosome–mRNA–(aminoacyl-tRNA)$_2$ arises. The two aminoacyl-tRNA's, one of which is the formylmethionyl-tRNA, are located in the ribosome side by side, in an immediate proximity, bound to two adjacent triplets.

Actually, whatever the real mechanism of the formation of the initial complex may be, three most important moments are evidently of the decisive importance for the initiation of translation: 1. the special properties of the formylmethionyl-tRNA determine its affinity for the peptidyl-tRNA-binding site and thereby the possibility of simultaneous attachment of two aminoacyl-tRNA's to the ribosome; 2. the formylmethionyl-tRNA recognizes and selects the initiating triplet on the template polynucleotide; 3. the recognition and selection of the initiating triplet by the formylmethionyl-tRNA molecule determines the proper setting of the phase in the subsequent reading of the template polynucleotide.

d) Formation of the First Peptide Bond

Thus, as a result of the binding of the initiator formylmethionyl-tRNA and the second, dependent on the next triplet of the template, aminoacyl-tRNA into the initial complex with the ribosome, there are all the prerequisites for the decisive event — the formation of the first peptide bond. Obviously the two tRNA's should be mutually orientated in such a way as to ensure a close arrangement of the groups that are to participate in the reaction of the peptide bond formation. Evidently this is automatically achieved due to appropriate mutual positions of the two tRNA-binding sites in the ribosome and the corresponding orientation of each tRNA molecule in its tRNA-binding site.

In all probability, peptide synthetase activity should be ascribed to the ribosome itself as such [TRAUT, MONRO, 1964; RYCHLIK, 1966 (1); ZAMIR et al., 1966; MONRO, 1967]; the strict mutual orientation of the tRNA molecules and appropriate drawing together of the corresponding groups that are to participate in the reaction should ensure spontaneous formation of the peptide bond in the ribosome.

The reaction of the peptide bond formation consists of substitution of $tRNA_{f\text{-}met}$ from its ester bond with the carboxyl of the formylmethionine residue by the free NH_2-group of the aminoacyl residue of the second aminoacyl-tRNA; the NH_2-group and the carboxyl form an amide (peptide) bond:

$$\text{CHO-NH}-\underset{}{\overset{(CH_2)_2-S-CH_3}{\underset{|}{}\text{CH}}}-\text{CO}\sim tRNA_{f\text{-}met}+NH_2-\overset{R}{\underset{|}{}\text{CH}}-\text{CO}\sim tRNA\rightarrow$$

$$\rightarrow \text{CHO}-\text{NH}-\overset{(CH_2)_2-S-CH_3}{\text{CH}}-\text{CO}-\text{NH}-\overset{R}{\text{CH}}-\text{CO}\sim tRNA+tRNA_{f\text{-}met}$$

Thus, as a result of the reaction between the two aminoacyl-tRNA's, dipeptidyl-tRNA and a free (deacylated) $tRNA_{f\text{-}met}$ are formed.

It can be imagined that during the reaction the activated carboxyl of the formylmethionine residue is transferred from its own $tRNA_{f\text{-}met}$ to the free amino group of the second aminoacyl-tRNA. Then as a result of the reaction, the tRNA residue of the dipeptidyl-tRNA molecule is found in the aminoacyl-tRNA-binding site, while the free $tRNA_{f\text{-}met}$ molecule is found in the peptidyl-tRNA-binding site (Fig. 19-5).

It was shown that it is the 50 S subparticle as such that bears the "enzymatic" (catalytical) center which performs the mentioned formylmethionyl transferase reaction, i.e., the formation of the peptide bond (MONRO, 1967). This center being a part of the peptidyl-tRNA-binding site is probably capable of holding the 3′-end sequence of the $tRNA_{f\text{-}met}$ with the adjoined aminoacyl residue (MONRO, MARCKER, 1967; MONRO, 1967; MONRO, VAZQUEZ, 1967). On the other hand, one may assume that this center possesses special affinity for the aminoacyl end (.... A-aminoacyl or ... CA-aminoacyl) of the aminoacyl-tRNA molecule (WALLER et al., 1966; RYCHLIK et al., 1967; MONRO, MARCKER, 1967; MONRO, VAZQUEZ, 1967). All this probably ensures an appropriate drawing together of the free amino group of the aminoacyl of the aminoacyl-tRNA molecule with the esterified carboxyl of the methionine of the formylmethionyl-tRNA molecule and the resulting from this the transfer of the formylmethionyl residue to the amino group of the aminoacyl-tRNA.

e) Initiation in Systems with the Participation of Synthetic Templates without Initiating Codons

If the template polynucleotide does not contain an initiating codon, the initiation of translation under physiological conditions is impossible or strongly hindered. As might be thought, the main cause lies in the fact that all the usual aminoacyl-tRNAs, being well retained in the aminoacyl-tRNA-binding site, are not retained in the peptidyl-tRNA-binding site; in such a case, the situation necessary for the formation of the first peptide bond occurs only occasionally and is of relatively low probability.

At the same time it is well known that in cell-free protein-synthesizing systems, the most varied set of synthetic template polynucleotides, containing no initiating codons, can be used for the synthesis of polypeptides. Consequently, in one way or another, the initiation of translation in experiments with synthetic polynucleotides can be obtained, i.e., it is possible to do without the initiating codon. The classical example of the synthesis of a polypeptide in a cell-free system with participation of a template without the initiating codon is the synthesis of polyphenylalanine in the system with polyU (NIRENBERG, MATTHAEI, 1961).

Just as it should have been expected, the initiation of polypeptide synthesis in such systems (for example, with polyU) possesses its own peculiarities and requires certain special conditions, deviating from the physiological. First of all, although the synthesis of a polypeptide with the participation of polynucleotides containing the initiating AUG or GUG codon begins immediately after the introduction of all the necessary factors into the system, there is, as a rule, a more or less prolonged *lag-period* ("induction period") in the system with polyU [NAKAMOTO et al., 1963; ALLENDE et al., 1964; NISHIZUKA, LIPMANN, 1966 (1)]. Only after a certain time has elapsed (from 30 sec to several minutes) does synthesis of the polypeptide begin in the system with polyU; it then increases rapidly and reaches the normal rate. This means that in a system with polyU, it is precisely initiation that is hindered, i.e., the bottleneck of the entire translation is the reaction of formation of the first peptide bond. But, as can be seen, despite the absence of the initiating codon, this reaction can be accomplished in the system, although at a very slow rate.

The above mentioned may be explained in the following way. As a result of the association of the ribosome with polyU, the binding of phenylalanyl-tRNA, and only phenylalanyl-tRNA, to the ribosome is permitted. The normal specific binding of the phenylalanine-tRNA occurs on the aminoacyl-tRNA-binding site. Entering of the phenylalanyl-tRNA on the peptidyl-tRNA-binding site is a rare and occasional event because of the low affinity of any normal aminoacyl-tRNA for this site. But still, since there is some small probability of the simultaneous presence of two properly orientated phenylalanyl-tRNA's on the ribosome, one in the aminoacyl-tRNA-binding site and the other in the peptidyl-tRNA-binding site, the first peptide bonds can be formed with a definite low frequency in the ribosomes of the system. As soon as diphenylalanyl-tRNA has been formed in the ribosome, the further course of translation encounters no difficulties and proceeds vigorously. Thus, gradually, during the lag period, ribosome after ribosome overcomes the indicated initiation barrier, and polypeptide synthesis in the system unfolds, reaching the normal rate.

An important special condition for polypeptide synthesis in a system with templates, containing no initiating codons, or in a system in the absence of formylmethionyl-tRNA, is a relatively high concentration of Mg^{++} ions, approximately twice as high as that used with natural mRNA's or synthetic templates containing initial AUG or GUG triplets in the presence of formylmethionyl-tRNA. It has been found that this increased concentration, 0.015 to 0.020 M Mg^{++}, is required only for the beginning of polypeptide synthesis, for the successful and rapid passage through the lag period, and when translation in the system is already proceeding fully, the

Mg^{++} concentration can be lowered to 0.007 – 0.010 M, which is more optimum for the entire process (REVEL, HIATT, 1965; NAKAMOTO, KOLAKOFSKY, 1966). From the well known fact that increased concentrations of Mg^{++} ions in the medium promote both a stronger association of the ribosome with the template and a stronger retention of tRNA on the ribosome, a simple explanation for the indicated "initiating role" of Mg^{++} suggests itself: at an increased concentration of Mg^{++}, the ternary complex of ribosome–template–aminoacyl-tRNA becomes sufficiently stable even without initiator tRNA, and at the same time, the affinity of the aminoacyl-tRNA for the peptidyl-tRNA-binding site is increased; from this the probability of an appropriate neighboring arrangement of the two aminoacyl-tRNA's in the ribosome increases, leading to the corresponding comparatively rapid formation of the first peptide bonds in the ribosomes of the system.

In cell-free systems, where template polynucleotides without initiation codons are used, an initiation of translation, however, can be achieved without increasing the Mg^{++} concentration and without overcoming the lag period. This, as has been found, can be done by adding exogenous peptidyl-tRNA to the system, where the specificity of the tRNA corresponds to the codons of the template polynucleotide. In such a case, the peptidyl-tRNA proves to be an effective initiator of the translation. For example, if diphenylalanyl-$tRNA_{phe}$ is added to a system with polyU, then no lag period is observed, translation begins immediately, and no increased concentration of Mg^{++} is required for initiation (NAKAMOTO, KOLAKOFSKY, 1966). Evidently exogenous peptidyl-tRNA can readily occupy the peptidyl-tRNA-binding site of the ribosome as a result of its great affinity for it [see above, Section II; RYCHLIK, 1966 (1, 2)]. As a result, the peptidyl-tRNA and the specifically bound aminoacyl-tRNA are found together, side by side on the ribosome, which permits immediate formation of a peptide bond — in this case between the peptidyl and the aminoacyl-tRNA. Thus, peptidyl-tRNA can replace the normal initiator (formylmethionyl)-tRNA in the initiation of translation.

2. Translation Proper ("Polymerization of Amino Acid Residues")

After the first peptide bond has been formed, and dipeptidyl-tRNA has appeared in the ribosome (Fig. 19-5), we can speak of the central stage of the functioning of the ribosome in the protein-synthesizing system — the stage of successive polymerization of amino acid residues into a polypeptide chain, or the stage of translation proper.

Although the basic points of the process of translation have already been discussed in one form or another, here it is advisable to briefly mention and summarize the general rules that lie at the basis of this process.

1. It is not free amino acids, but aminoacyl-tRNA's that participate in the polymerization of amino acid residues into a polypeptide chain. Correspondingly, the template polynucleotide, strictly speaking, is not a template for the polymerization of amino acids, but only a linear sequence of nucleotides, setting a definite order of the different aminoacyl-tRNA's, one following another, in the synthesis of the polypeptide chain.

2. The process of synthesis of the polypeptide chain is not an instantaneous polymerization of aminoacyl residues as a result of the array of numerous aminoacyl-

tRNA's on the template, but represents a stepwise, strictly sequential growth of the polypeptide chain, one amino acid residue at a time. Consequently, the template polynucleotide does not set the spatial order of arrangement of the different aminoacyl-tRNA's but the temporal order of addition of their aminoacyl residues to the growing polypeptide.

3. During the entire process of polymerization, the growing polypeptide is retained by the ribosome. The addition of each successive aminoacyl residue occurs at the C-end of the polypeptide, i.e., it is the C-end that is growing (Bishop et al., 1960; Dintzis, 1961; Rychlik, Šorm, 1962). The tRNA having brought the last added aminoacyl residue, remains bonded to it, i.e., throughout the entire process of translation the C-end of the growing polypeptide is never free, but is bound by an ester bond to tRNA [Nathans, Lipmann, 1961; Takanami, 1962; Gilbert, 1963 (2); Bretscher, 1963, 1965; Rychlik, 1965]. The addition of the next aminoacyl residue consists of the substitution of tRNA by aminoacyl-tRNA. Thus, each event of peptide bond formation is a reaction between peptidyl-tRNA′ and aminoacyl-tRNA″, resulting in the formation of peptidyl-tRNA″ with its peptide residue increased by one link, and free tRNA′:

$$\text{peptidyl(n)-tRNA}' + \text{aminoacyl-tRNA}'' \rightarrow \text{peptidyl(n+1)-tRNA}'' + \text{tRNA}'.$$

4. Each such event of formation of a peptide bond is a cycle made up of several successive steps: 1. the entering of aminoacyl-tRNA″ into the ribosome and its appropriate orientation alongside the peptidyl-tRNA′; 2. transfer of the C-end of the peptide residue from tRNA′ to the free NH_2 group of the aminoacyl residue of aminoacyl-tRNA″; 3. displacement of the formed peptidyl(n+1)-tRNA″ to the site which the peptidyl(n)-tRNA′ occupied at the beginning of the cycle. Probably a displacement of the chain of the template polynucleotide relative to the ribosome by one triplet in the direction from the 5′-end to the 3′-end is accomplished simultaneously with step 3.

5. The entire process of the polypeptide chain synthesis is made up of numerous (according to the number of amino acid residues in the polypeptide) repetitions of such cycles. Consequently, during the process of translation the ribosome represents a cyclically operating machine; in each cycle the formation of one peptide bond and the drawing over of the template polynucleotide by one triplet is accomplished.

Since translation proper, or polymerization, on the whole is reduced to numerous *repetitions* of the same cycle, a more detailed consideration of the latter is necessary and sufficient for the description and characterization of the entire process. For a consideration of the cycle its individual participants will first be described with the following general picture of their interaction during the functioning of the ribosome.

a) Peptidyl-tRNA and Its Retention in the Translating Ribosome

During the process of initiation of translation, formylmethionyl-$tRNA_{f\text{-}met}$ reacts with a second aminoacyl-tRNA, for example, with alanyl-$tRNA_{ala}$, so that formylmethionyl-alanyl-$tRNA_{ala}$ and free $tRNA_{f\text{-}met}$ are formed. The formed formylmethionyl-alanyl-$tRNA_{ala}$ means the appearance of a peptidyl-tRNA in the ribosome, in this case a dipeptidyl-tRNA. Subsequently the peptidyl residue is lengthened on account of analogous sequential reactions, for example:

formylmethionyl-alanyl-$tRNA_{ala}$ + seryl-$tRNA_{ser}$ → formylmethionyl-alanyl-seryl-$tRNA_{ser}$ + $tRNA_{ala}$; etc.

Thus, on the growing end of the peptide there is always a tRNA covalently bound to it. In other words, if translation has begun, then during its entire course, peptidyl-tRNA will be permanently present on the ribosome; the peptidyl residue is successively lengthened in the process, while tRNA residues are correspondingly replaced each time.

The beginning of translation, i.e., the formation of peptidyl-tRNA, irreversibly consolidates the association of the ribosome with the template. Under physiological conditions, the translating ribosome, as long as it has peptidyl-tRNA, evidently cannot dissociate from the template polynucleotide. Consequently, in all probability it is precisely peptidyl-tRNA that is responsible for the irreversible retention of the ribosome on the template polynucleotide. More likely this is somehow connected with the fact that peptidyl-tRNA itself is very firmly retained by the ribosome. There is no question of it being exchanged with exogenous derivatives of any tRNA's. The association is so strong that the ribosome can be dissociated into subparticles, by means of lowering the Mg^{++} concentration to 10^{-4} M but the peptidyl-tRNA is not released and remains in a complex with the 50 S subparticle [GILBERT, 1963 (2)]. It seems that it is only by destruction of the 50 S subparticle [GILBERT, 1963 (2)] or by its incubation in magnesium-free medium (BRESLER et al., 1966), the latter leading to a loosening of its compactness, that peptidyl-tRNA can be released.

Such firm retention of peptidyl-tRNA in the translating ribosome is evidently due to the cooperative effect of binding both of the tRNA residue itself and of its peptidyl residue. When the peptide is removed from tRNA, they both can no longer be firmly retained on the ribosome or in association with its 50 S subparticle, and are easily released into solution [GILBERT, 1963 (2); TRAUT, MONRO, 1964; CANNON, 1967]. At the same time, for the firm retention indicated above, evidently dipeptidyl-tRNA is sufficient (CANNON, 1967).

From the fact that after dissociation of the translating ribosome into subparticles, the peptidyl-tRNA remains firmly bound to the 50 S subparticle, it can be concluded that the site for the binding of peptidyl-tRNA is localized precisely on the 50 S subparticle of the ribosome. It has already been noted (Section II, 3, c) that there is a tRNA-binding site here, but the free tRNA's and aminoacyl-tRNA's exhibit only a slight affinity for it: binding, if it does occur, is very labile, requires a Mg^{++} concentration no less than 0.01 M, and the bound tRNA's or aminoacyl-tRNA's exist in a dynamic equilibrium with exogenous tRNA's, readily and rapidly exchanging with them (CANNON, 1967). It has also been indicated (Section III, 1) that the initiator formylmethionyl-tRNA possesses an appreciably greater, in comparison with the usual tRNA's, affinity for this tRNA-binding site on the 50 S subparticle. Finally, it has been noted that in all probability, precisely this site is responsible for the firm retention of peptidyl-tRNA in the ribosome. In both the latter cases the affinity increases sharply in comparison with aminoacyl-tRNA, evidently as a result of the appearance of a supplementary point of binding at the peptide group closest to the tRNA. This site is denoted as the "peptidyl-tRNA-binding site" of the ribosome.

Thus, the first necessary participant, or component, of any cycle during translation is peptidyl-tRNA. It is firmly retained on the 50 S subparticle of the ribosome, occupying its peptidyl-tRNA-binding site.

b) Entering of Aminoacyl-tRNA into the Ribosome

The successive growth of the polypeptide chain during translation presupposes successive entering of aminoacyl-tRNAs into the ribosome. To ensure strict determination of the amino acid sequence of the polypeptide, the specificity of the entering aminoacyl-tRNA must correspond unambiguously to the codon situated in the ribosome at the given moment. Thus, a rigorous requirement is set for the entering of aminoacyl-tRNA into the translating ribosome: no aminoacyl-tRNA other than the aminoacyl-tRNA possessing an anticodon complementary to the codon of the template can normally enter the working ribosome. Consequently, the functioning ribosome, with the participation of the template polynucleotide, primarily performs a strict *selection* of a definite specific aminoacyl-tRNA from the medium.

It has already been indicated that the function of selection of the aminoacyl-tRNA is retained by the isolated 30 S subparticle, associated with a template polynucleotide or oligonucleotide (Section II, 2, d). Most likely it is the 30 S subparticle, bound at each given moment to a definite codon, that is entirely responsible for the function of selection in the translating ribosome. It has been determined that a special aminoacyl-tRNA-binding site is localized on the 30 S subparticle; this site itself apparently exhibits some affinity for tRNA, but insufficient to retain it; however, if the complementary interaction between the anticodon of tRNA and the codon of the template polynucleotide or oligonucleotide associated with the ribosome is added to this affinity, then the corresponding tRNA or aminoacyl-tRNA is retained by this tRNA-binding site. It is just this that may be the basic mechanism of the selection of specific aminoacyl-tRNA during translation.

The conditions of specific binding of aminoacyl-tRNA's with the aminoacyl-tRNA-binding site on the 30 S subparticle of the ribosome were analyzed in detail above (Section II, 2, a).

For the addition of an aminoacyl residue to a peptide chain to take place, the specific selection of aminoacyl-tRNA from the medium and its entering into the ribosome must also be followed by a quite definite, rigid orientation of the aminoacyl-tRNA molecule relative to the peptidyl-tRNA firmly attached to the peptidyl-tRNA-binding site. This orientation must provide an appropriate drawing together of the amino group of the aminoacyl residue of aminoacyl-tRNA to the ester group with which the peptidyl is bonded to tRNA in the peptidyl-tRNA molecule.

Inasmuch as complementary interaction with the *adjacent* codons of the template polynucleotide is realized for both the tRNA residues, the aminoacyl-tRNA molecule, as is evident, must be situated in close proximity, side by side, with the peptidyl-tRNA molecule over a substantial extent of their length. Then, the aminoacyl-tRNA-binding site of the 30 S subparticle must be immediately adjacent to the peptidyl-tRNA-binding site of the 50 S subparticle where the peptidyl-tRNA is firmly retained. Hence, both the tRNA-binding sites must be localized at the very boundary between the subparticles of the ribosome, most likely on the very contacting surfaces of the subparticles.

Moreover, in all probability the aminoacyl end of the aminoacyl-tRNA-molecule is not attached to the aminoacyl-tRNA-binding site of the 30 S subparticle, but "hangs" from it. At the same time there is a special center located on the peptidyl-tRNA-binding site of the 50 S subparticle in the region of the ester group connecting the peptidyl with the tRNA and possessing an affinity for the aminoacyl end of the aminoacyl-tRNA molecule (MONRO, 1967; MONRO, VAZQUEZ, 1967). It may be assumed that this "peptidyl-transferase center" actually possesses the necessary orientating function: it binds the aminoacyl end of the aminoacyl-tRNA molecule alongside the ester group of the peptidyl-tRNA molecule.

Thus, as a result of the binding and orientation of the aminoacyl-tRNA molecule in the ribosome, it is located mainly on the 30 S subparticle, but the aminoacyl end spreads out to the 50 S subparticle and, on the contrary, the peptidyl-tRNA is bound to the 50 S subparticle but its anticodon interacts with the mRNA attached to the 30 S subparticle.

c) Formation of a Peptide Bond

In all probability, it is the retention and proper orientation of the aminoacyl-tRNA molecule next to the peptidyl-tRNA that creates the possibility of the formation of a peptide bond in the ribosome. For this to happen first of all, as was mentioned above, the free amino group of the aminoacyl residue of the aminoacyl-tRNA must be close to the esterified carboxyl of the peptidyl residue of peptidyl-tRNA. Due to the presence of the peptidyl-transferase center on the 50 S subparticle of the ribosome [TRAUT, MONRO, 1964; RYCHLIK, 1966 (1); ZAMIR et al., 1966; MONRO, 1697] a covalent bond arises between the amino group and the carboxyl, as a result of which the tRNA, that was connected by the ester bond with the carboxyl of the peptide, is displaced from this bond. Thus, the reaction represents a substitution of the tRNA residue by the aminoacyl-tRNA residue; the ester bond is replaced by the amide (peptide) bond.

$$\mathrm{HCO{-}NH{-}\underset{|}{\overset{(CH_2)_2-S-CH_3}{CH}}{-}CO{-}[NH{-}\overset{R}{CH}{-}CO{-}]_n{-}NH{-}\overset{R'}{CH}{-}CO{\sim}tRNA'} +$$

$$+\ \mathrm{NH_2{-}\underset{R''}{CH}{-}CO{\sim}tRNA''} \rightarrow$$

$$\rightarrow \mathrm{HCO{-}NH{-}\overset{(CH_2)_2-S-CH_3}{CH}{-}CO{-}[NH{-}\overset{R}{CH}{-}CO{-}]_n{-}NH{-}\overset{R'}{CH}{-}CO{-}NH{-}\overset{R''}{CH}{-}CO{\sim}tRNA''} +$$

$$+\ \mathrm{tRNA'}$$

The sense of the reaction is that the peptide is lengthened by one aminoacyl residue. As can be seen, the C-end of the peptide grows.

d) Translocation of the tRNA Residue and Displacement of mRNA Chain

As a result of the reaction, an unstable situation is created in the ribosome: the tRNA', deprived of its peptide, now does not possess such a strong affinity for its tRNA-binding site on the ribosome (the peptidyl-tRNA-binding site of the 50 S subparticle) and thus can no longer be firmly retained in it; on the other hand, tRNA'', which is now bound to the peptide must, like any peptidyl-tRNA, acquire a strong affinity for the peptidyl-tRNA-binding site. It may be that simply on account of the fact that the affinity of the peptidyl-tRNA'' for the peptidyl-tRNA-binding site substantially exceeds its affinity for the aminoacyl-tRNA-binding site, where tRNA'' was present at the moment of the reaction, it might be spontaneously transferred to the peptidyl-tRNA-binding site, with a displacement of the weakly retained free (deacylated) tRNA' from it.

However, tRNA'' is complementarily bound with its anticodon to the codon of the template polynucleotide; in such a situation, as has been indicated, a rather reliable retention of the tRNA residue in the aminoacyl-tRNA-binding site is ensured. Then, the tRNA', situated in the peptidyl-tRNA-binding site, is also bound with its anticodon to the template codon complementary to it; this codon is "alien" to the tRNA'', which is ready to pass over to the peptidyl-tRNA-binding site. Finally, if the tRNA'' with its peptide were simply to pass over to the peptidyl-tRNA-binding site and thereby free the aminoacyl-tRNA-binding site where it was situated hitherto, then the old codon situated there, complementary to the anticodon of tRNA'' would prevent the entering into the ribosome of a next aminoacyl-tRNA of any different specificity. Consequently, the passing over of the peptidyl-tRNA'' from the aminoacyl-tRNA-binding site to the peptidyl-tRNA-binding site must occur together with a shift of the corresponding codon from one site to the other, i.e., simultaneously and synchronously with the shift of the template polynucleotide relative to the ribosome by one triplet.

It is believed that precisely the process of shifting of the template polynucleotide and tRNA residue, immediately following the formation of the peptide bond, requires energy at the expense of the GTP cleavage, i.e., requires the presence of GTP and the protein G-factor [Noll et al., 1963; Traut, Monro, 1964; Conway, Lipmann, 1964; Nishizuka, Lipmann, 1966 (2)].

Two theoretically possible manners of the drawing over (shifting) of the template polynucleotide relative to the ribosome can be conceived.

1. The shifting of the template is an active, *driving* process with respect to the process of displacement of the tRNA residue of the peptidyl-tRNA molecule from one site to the other. This means that the formation of the peptide bond in some uncomprehended way induces a local conformational rearrangement of the ribosome (possibly the mRNA-binding site and its neighborhood), which shifts the template by three nucleotide residues (about 10 Å). As a result of the shift of the template, the tRNA residue of the peptidyl-tRNA molecule is drawn over by its codon to the peptidyl-tRNA-binding site, for which the peptidyl-tRNA already possesses increased affinity, while the free tRNA that was there is readily displaced. For the indicated rearrangement (transconformation) to displace the template, energy should be required which can be supplied by cleavage of a GTP molecule, i.e., the rearrangement should be a GTP-dependent process. On the other hand, rearrangement may be spontaneous in response to formation of a peptide bond, and then relaxation should be GTP-dependent.

2. The shifting of the template is a passive, *driven* process with respect to the process of displacement of the tRNA residue of the peptidyl-tRNA molecule from one tRNA-binding site to the other. This means that the tRNA residue of the peptidyl-tRNA molecule spontaneously, or on account of the energy of GTP cleavage, passes

over to the site on the 50 S subparticle and draws along the codon bound to it, thereby leading to a shift of the template by one triplet. This transfer of the tRNA residue from the aminoacyl-tRNA-binding site to the peptidyl-tRNA-binding site can directly be induced by the peptide bond formation, since the given tRNA residue becomes then a part of the peptidyl-tRNA molecule and thus must acquire an increased affinity for the peptidyl-tRNA-binding site. Such a representation of the shifting of the template as a driven process gives a simpler and therefore very attractive explanation of a mechanism of translocation, the strict conjugation of the process with the formation of a peptide bond and the always errorless displacement of the template presicely by a triplet, no more and no less. Hence this second manner seems to merit more attention as a working hypothesis.

e) General Scheme of the Working Cycle of Translation

The very fact that the peptidyl-tRNA and the aminoacyl-tRNA can be simultaneously located on the functioning ribosome led to the postulating of the schemes of the functioning of the ribosome, where it was claimed that at least two separate fixed tRNA-binding sites, the aminoacyl-tRNA-binding site and the peptidyl-tRNA-binding site existed (WATSON, 1964, 1965; WARNER, RICH, 1964; WETTSTEIN, NOLL, 1965). Subsequently more direct experimental evidence was obtained attesting that the complete ribosome actually bears two tRNA-binding sites with different properties (see above — Section II, and also III, 2, a, b). In the suggested schemes, and also in the earlier schemes of different authors [LIPMANN, 1963; GILBERT, 1963 (2); TRAUT, MONRO, 1964], it was also postulated that the peptide bond formation in the ribosome occurs as the transfer of carboxyl of the peptidyl residue from the tRNA to the amino group of the aminoacyl-tRNA. Finally, the third most important postulate of the schemes suggested was the assertion that immediately after the formation of the peptide bond the tRNA residue of the newly formed peptidyl-tRNA is displaced from the aminoacyl-tRNA-binding site to the peptidyl-tRNA-binding site. J. Watson's scheme (WATSON, 1964, 1965) was the first to summarise and formulate all these ideas in the best and coherent manner.

The following schematical description of the sequence of events in a single working cycle of the translating ribosome can be given (Fig. 20) based on the main assumptions of J. Watson's scheme and on the strength of all the experimental evidence cited above (Section III, a—d). Let us begin at some arbitrary moment characterized by the presence of the peptidyl-tRNA in the peptidyl-tRNA-binding site of the 50 S subparticle and of the aminoacyl-tRNA in the aminoacyl-tRNA-binding site of the 30 S subparticle; both tRNA's occupy two adjacent codons of the mRNA chain (Fig. 20, top right). There is a peptidyl-transferase center on the 50 S subparticle which is evidently an integral part of the peptidyl-tRNA-binding site. This center retains the 3'-end nucleotide sequence and the adjacent aminoacyl residue of the peptidyl-tRNA molecule (MONRO, MARCKER, 1967; MONRO, 1967). In addition the peptidyl-transferase center of the 50 S subparticle must evidently have an affinity for the aminoacyl end of the aminoacyl-tRNA molecule (probably for the terminal combination ...A-aminoacyl or ..CA-aminoacyl; see WALLER et al., 1966; RYCHLIK et al., 1967, in conjunction with the data of MONRO, 1967; MONRO, VAZQUEZ, 1967). *Step 1* (Fig. 20-1): the aminoacyl end of the amino acyl-tRNA molecule bound to the 30 S subparticle is positioned and orientated immediately adjacent to the esterified carboxyl of the peptidyl in the peptidyl-transferase center of the 50 S subparticle. *Step 2* (Fig. 20-2): the carboxyl of the peptidyl is transferred from the ribose hydroxyl of its tRNA to the amino group of the aminoacyl of the aminoacyl-tRNA, i.e., a peptide bond is formed between the C-end of the peptidyl and the amino group of the aminoacyl-tRNA; deacylated tRNA is left in the peptidyl-tRNA-binding site. *Step 3* (translocation) (Fig. 20-3): the tRNA residue of the newly formed peptidyl-tRNA molecule is displaced from the amino-acyl-tRNA-binding site on the 30 S subparticle to the peptidyl-tRNA-binding site on the 50 S subparticle, drawing along the codon of the template polynucleotide bound

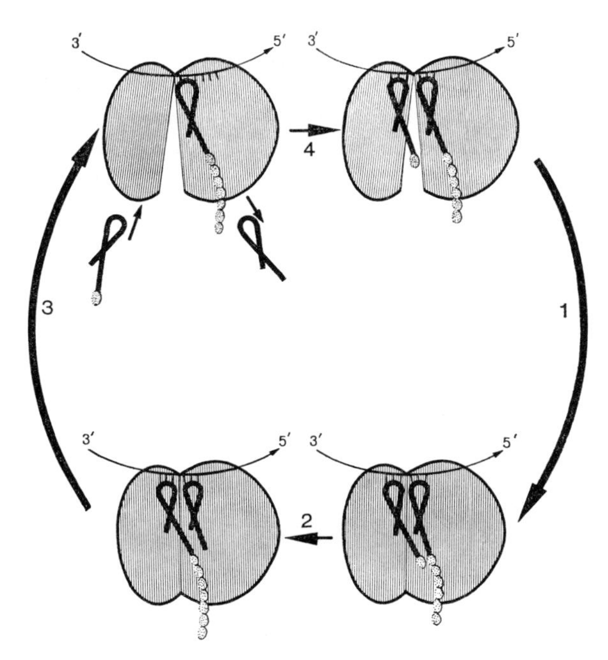

Fig. 20. Probable scheme of the operating cycle of the ribosome

to it and replacing the deacylated tRNA on the peptidyl-tRNA-binding site. It may be assumed that this translocation is brought about by some conformational change of the ribosome occurring at the expense of energy of the GTP molecule cleavage. *Step 4* (Fig. 20-4): after the translocation has taken place, the freed aminoacyl-tRNA-binding site with the subsequent vacant codon positioned on it specifically binds a new aminoacyl-tRNA corresponding to this codon. As a result, the system returns to the position from which the consideration of the cycle commenced.

Thus, as a result of a complete single cycle, a) one peptide bond is synthesized — the polypeptide is lengthened by one amino acid; b) the template polynucleotide is shifted by one triplet relative to the ribosome — one codon is read out. It is the repetition of these cycles that creates the process of translation, as a result of which the template polynucleotide is read out, and a polypeptide chain is synchronously synthesized. It may be recalled here that the template polynucleotide is shifted (read out) in the direction from the 5′-end to the 3′-end. The polypeptide chain, as can be seen from the cycle described, grows with its C-end, i.e., is synthesized from the N-end to the C-end.

f) On the Mechanism of the Working of the Ribosome: A Hypothesis on the Periodical Locking and Unlocking of the Ribosome

Although at present the probable sequence of events during the functioning of the ribosome can, in the form of a scheme, be *described* (see preceding Section III, 2, e), up to now no concrete suggestions have been put forward concerning the molecular mechanism which could *explain* this sequence of events, array these events into acts logically ensuing one from another and connect them into a unified system. In other words, up to the present no theory has been presented on the functioning of the ribosome. Moreover, to date, all considerations regarding the functioning of the ribosome bore a purely abstract nature and were in no way connected with the structure of the ribosome, in particular with its most peculiar structural feature — its construction from two unequal subparticles.

The present subsection is devoted to the exposition of a hypothesis (Spirin, 1968) where the principle of a unified driving mechanism providing for all the known spatial displacements in the functioning process of the ribosome is postulated. This mechanism straightforwardly explains and connects into a system all the known acts of the working cycle of the ribosome: the entering of aminoacyl-tRNA, the release of deacylated tRNA, the formation of peptide bonds, the translocation of both the mRNA and the peptidyl-tRNA and the cleavage of GTP.

The first prelusive fact for the hypothesis is that the functioning ribosome is necessarily constructed from two unequal subparticles. The second is that the ribosome has separate aminoacyl-tRNA-binding and peptidyl-tRNA-binding sites.

The initial postulates on which the main statement of the hypothesis is based are the following assertions.

Postulate 1: the aminoacyl-tRNA-binding site and the peptidyl-tRNA-binding site are localized on two different subparticles of the ribosome. Practically this postulate is taken for granted throughout the preceeding exposition. Indeed, there are numerous indications that the mRNA-binding center and the aminoacyl-tRNA-binding site evidently including in itself the mRNA-binding center are localized on the small (30 S) subparticle of the ribosome (see Section II, 1c and 2d). On the other hand it can be deduced from the available data [Gilbert, 1963 (2); Cannon, 1966; Traut, Monro, 1964; Monro, 1967] that the peptidyl-tRNA-binding site in the ribosome is localized on the large (50 S) subparticle (see Section II, 3c, 4, and Section III, 2a).

Postulate 2: the place of localization of the mRNA-binding center and the tRNA-binding sites are the contacting (internal, facing each other) surfaces of the subparticles. Such a localization of tRNA-binding sites means that the peptidyl-tRNA and the aminoacyl-tRNA are situated *within* the ribosome, held between its subparticles and that the events of peptide bond synthesis in the functioning ribosome also take place within the ribosome on the border between the subparticles. Some indirect evidence can be indicated in favor of this concept: 1. If the dissociation of subparticles is prevented by fixing the ribosomes with formaldehyde, then the template polynucleotide is not released from the ribosome when the Mg^{++} concentration is lowered; at the same time the template polynucleotide is readily released from the 30 S subparticle–template complex treated with formaldehyde when the Mg^{++} concentration is lowered (MOORE, 1966). 2. The presence of peptidyl-tRNA in the ribosome stabilizes the association between the subparticles (SCHLESSINGER et al., 1967). 3. A rather long C-fragment of the growing peptide in the ribosome is found to be protected from exogenous protease action (MALKIN, RICH, 1967). Of course, these are very weak arguments and the proposed supposition should be accepted only as a postulate.

Both above mentioned postulates, 1 and 2, are demonstrated schematically in Fig. 18.

Postulate 3: the complete functioning ribosome (70 S) can be in two different states, locked (the subparticles tightly associated) and unlocked (the subparticles in a loose association or slightly drawn apart). As the peptidyl-tRNA and the aminoacyl-tRNA are located on different ribosomal subparticles (postulate 1), then the subparticles must be tightly drawn together to form a peptide bond. At the same time, inasmuch as both tRNA-binding sites are localized on the contacting surfaces of the subparticles (postulate 2), the subparticles must be slightly drawn apart to allow the entry of aminoacyl-tRNA into the ribosome and the release of deacylated tRNA from the ribosome. The above supposition conforms with observations that not just an association of two subparticles, but an association of a strictly definite strength, neither too small nor too great, is necessary for the functioning of the ribosome (see part "Structure of the Ribosome", Section IV, 1).

The main statement of the hypothesis: the periodical locking and unlocking of the ribosome (drawing together and slight drawing apart of the subparticles) is the driving mechanism providing for all the spatial displacements of tRNA and mRNA in the process of translation.

To begin with, let us consider such a state of the functioning ribosome when the aminoacyl-tRNA is between the subparticles, bound by its anticodon to the mRNA codon and retained by the contacting surface of the 30 S subparticle, while the peptidyl-tRNA is retained by the contacting surface of the 50 S subparticle (Fig. 20, top right). If the subparticles in such a ribosome are locked tightly (closely associated) (Fig. 20-1) then the aminoacyl-tRNA and the peptidyl-tRNA are alongside each other, ensuring a close contact of the amino group of aminoacyl with the ester group of peptidyl; since there is a peptidyl-transferase center on the 50 S subparticle, a peptide bond between the amino group of aminoacyl-tRNA and the C-end of peptidyl arises (Fig. 20-2). Thus, after the formation of a peptide bond in the locked ribosome there is the following situation: a) the tRNA in the peptidyl-tRNA-binding site is deacylated, free; the retention of such a tRNA in the peptidyl-tRNA-binding site must be very weak, but inasmuch as the subparticles are closely associated (the ribosome is locked) it cannot leave this site and emerge from the ribosome; b) the tRNA residue in the aminoacyl-tRNA-binding site also, as before, occupies its place on the 30 S subparticle although this residue has now become a constituent part of the peptidyl-tRNA, therefore acquiring an increased affinity to the peptidyl-tRNA-binding site; c) the peptidyl residue is retained on the 50 S subparticle.

Now let us draw the subparticles apart (Fig. 20-3). Inasmuch as the end of the tRNA residue is attached with its peptidyl to the 50 S subparticle, the drawing apart of the subparticles must lead to the *drawing out* of the tRNA residue from the aminoacyl-tRNA-binding site on the 30 S subparticle. Then the corresponding triplet of mRNA associated with the tRNA, will also be drawn out from the aminoacyl-tRNA-binding site of the 30 S subparticle together with the tRNA residue. At the same time the drawing apart of the subparticles will allow the weakly retained deacylated tRNA to emerge from the ribosome, leading to the vacation of the peptidyl-tRNA-binding site on the 50 S subparticle.

The drawn out tRNA residue from the 30 S subparticle must then settle on the peptidyl-tRNA-binding site of the 50 S subparticle, and the mRNA codon bound to it correspondingly moves along the 30 S subparticle. It is just this that means the translocation of the tRNA residue of the peptidyl-tRNA molecule with the simultaneous drawing over of the corresponding mRNA triplet. The drawing out of this triplet from the aminoacyl-tRNA-binding site of the 30 S subparticle will lead to the positioning on this site of the following adjacent mRNA triplet which means a shift of the mRNA chain by one triplet relative to the ribosome. Inasmuch as the particles are now drawn apart, the corresponding aminoacyl-tRNA from the surrounding medium will immediately settle on the freed aminoacyl-tRNA-binding site of the 30 S subparticle with the new vacant codon (Fig. 20-4).

The described cycle may be further repeated, again drawing the subparticles together, then drawing them apart. The transfer of the peptidyl C-end from the tRNA residue to the aminoacyl-tRNA residue will occur each time in the ribosome with the subparticles drawn together, while the drawing apart will lead to the translocation of the tRNA residue and the driven by it shift of the mRNA chain by one triplet, as well as permit the release of the deacylated tRNA from the ribosome and the entry of new aminoacyl-tRNA. Thus, there is on hand a mechanism where periodic cyclic movement (the drawing together and apart of the subparticles) is transformed into polar translational movement (the shift of the mRNA chain); this same cyclic movement provides for the successive linear growth of the peptide chain.

It is evident that namely the entry of aminoacyl-tRNA onto the aminoacyl-tRNA-binding site of the 30 S subparticle must induce the locking of the ribosome (drawing together of the subparticles). It is most probable that the locking is stipulated by the affinity of the aminoacyl end of the aminoacyl-tRNA molecule to the peptidyl-transferase center of the 50 S subparticle (see Section III, 2, b-c). In this case the entry of deacylated tRNA should not induce the locking. This is in full agreement with data that aminoacyl-tRNA, associating with the ribosome in the presence of a template, is no longer displaced by exogenous tRNA's of the same specificity (does not exchange with them), whereas bound deacylated tRNA is easily displaced by both deacylated and acylated exogenous tRNA's (NIRENBERG, LEDER, 1964; SEEDS, CONWAY, 1966).

The reverse process, the drawing apart of the subparticles (unlocking of the ribosome), probably requires an expenditure of energy. A known source of energy for the functioning of the ribosome is the cleavage of GTP into GDP and inorganic phosphate. Most likely this energy of GTP cleavage is applied mainly to the drawing apart (unlocking) of the subparticles. It is just here, by this drawing apart, that the work of drawing out the tRNA residue with the corresponding mRNA codon from the aminoacyl-tRNA-binding site of the 30 S subparticle is done. In agreement with this assumption are reports that GTP is necessary for the complete reacting of peptidyl-tRNA bound within the ribosome with puromycin (in the absence of GTP only half of the peptides are transferred to the puromycin) (TRAUT, MONRO, 1964; see lower, "Appendix"). Numerous other experimental facts, and in particular data that GTP reverses the inhibition of the translation by deacylated tRNA (SEEDS, CONWAY, 1966), the fact that the cleavage of GTP is necessary as a preliminary act for the reaction of formylmethionyl-tRNA with puromycin (HERSHEY, THACH, 1967; OHTA et al., 1967) and for the formation of the first peptide bond in the initiation of translation (THACH et al., 1967), and some other facts can be readily interpreted within the framework of the proposed hypothesis (SPIRIN, 1968).

3. Termination of Translation

Hence, according to the mechanism of translation, the growing polypeptide chain in the ribosome always carries a tRNA residue on its C-end. Due to the presence of tRNA, the growing polypeptide chain is firmly retained in the ribosome. Completion of the synthesis of the polypeptide chain of the protein must consequently mean its release from the bond with the last tRNA and from the ribosome.

It has been found that the reading out of the template polynucleotide up to the 3′-end itself does not ensure release of the peptide from tRNA and from its association with the ribosome. Thus, the polymerization of amino acids using many synthetic template polynucleotides in cell-free systems does not at all end in the formation of free polypeptides, but gives as a result only the corresponding peptidyl-tRNA's associated with the ribosome [GILBERT, 1963 (2); TAKANAMI, YAN, 1965; THACH et al., 1965; BRETSCHER et al., 1965; GANOZA, NAKAMOTO, 1966].

On the other hand, when natural polycistronic mRNA's are used, the synthesized polypeptide chains of proteins are released after each cistron is read out, i.e., after passage through some internal regions of the template polynucleotide, independent of its 3′-end [OHTAKA, SPIEGELMAN, 1963; CAPECCHI, 1966 (2)].

The enumerated facts could not but lead to the assumption that the release of the polypeptide from its bond with tRNA and from the ribosome is somehow induced by a special nucleotide combination within the chain of the template polynucleotide. Such combinations must be located at the end of each cistron of natural mRNA's. These nucleotide combinations have been called termination codons.

a) Termination Codons

In experiments using synthetic template polynucleotides in cell-free systems of *E. coli*, it has been noted that in contrast to most of the synthetic polynucleotides, the random copolymer poly(U,A) leads to the formation of a substantial fraction of free peptides (TAKANAMI, YAN, 1965; BRETSCHER et al., 1965; GANOZA, NAKAMOTO, 1966). The fraction of released peptides depends upon the ratio of A and U in the polynucleotide: it is greater when A predominates in the polymer. This permitted to assume that a triplet consisting of two A and one U may be a terminating codon. Free peptides were formed with substantial efficiency when the random template polynucleotide poly(A,U,I) was used as well (TAKANAMI, YAN, 1965). Since I plays the role of an equivalent of G in the template it might be concluded that the combination of A, U, and G in the template polynucleotide can also perform a terminating function.

In the testing of the most varied synthetic trinucleotides for binding of aminoacyl-tRNA in the cell-free system (NIRENBERG, LEDER, 1964), it was found that only a few triplets do not bind any of the usual aminoacyl-tRNA's, including the triplets UAA, UAG, and UGA (NIRENBERG et al., 1965; BRIMACOMBE et al., 1965). Subsequently it was definitively shown on systems with synthetic templates that only these three triplets among all the 64, UAA, UAG, and UGA, are actually "nonsense" triplets, i.e., triplets that do not code any of the amino acids (see KHORANA et al., 1966; MORGAN et al., 1966). It is probably to these triplets that the terminating function was ascribed.

Recently the terminating function of the UAA triplet was directly demonstrated in experiments *in vitro*, where the synthetic polynucleotides $AUGUA_n$, $AUGU_2A_n$, $AUGU_3A_n$ and $AUGU_4A_n$ were employed in the role of the templates in the protein-synthesizing cell-free system (LAST et al., 1967): in the first case no peptides were synthesized, in the second the formylmethionyl-leucyl-polylysine was synthesized, in the third the formylmethionyl-phenylalanyl-polylysine, and in the fourth case (AUGUUUUAA...) only the dipeptide formylmethionyl-phenylalanine was formed and at this point the translation terminated.

Somewhat earlier than all the cited biochemical experiments the genetic analysis of the nonsense "amber" and "ochre" mutations gave an evidence that the triplets UAA and UAG possess a terminating function in mRNA of *E. coli* (BRENNER et al., 1965).

In the case of the "amber" mutation in a definite gene, just as in the case of most of the usual point mutations, a given functional protein is not produced in the cells. However, the usual point mutations, as is well known, lead to the synthesis of an inactive analog of this protein, i.e., its complete polypeptide chain with one amino acid replaced. Instead, "amber" mutations lead to the accumulation of uncompleted N-terminal fragments of the polypeptide chains of this protein (SARABHAI et al., 1964; STRETTON, BRENNER, 1965). Consequently, the "amber" mutation consists not of replacement of the amino acid residue, like most of the usual point mutations, but of an *interruption* of the polypeptide chain at the point corresponding to the changed codon of the given gene.

Hence, the changed (mutated) codon in "amber" mutants does not determine any amino acid, but is "nonsense" and determines cessation or termination of the translation process. It might be thought that the "amber" mutation thus means a change of some sense codon into a termination codon.

In one group of the investigated "amber" mutants, a polypeptide chain is interrupted at a point where a glutamine residue should normally be [STRETTON, BRENNER, 1965; NOTANI et al., 1965; WEIGERT, GAREN, 1965 (1)], while in others it is interrupted at a site where tryptophan is in the normal protein [WEIGERT, GAREN, 1965 (1); BRENNER et al., 1965]. Consequently, the "amber" mutation consists of a change either of the codon of glutamine or of the codon of tryptophan into the termination codon. Evidently the only possible codon that might originate from one point mutation both of the glutamine codon (CAA, CAG) and of the tryptophan codon (UGG) can be the triplet UAG (CAG $\rightarrow$ UAG or UGG $\rightarrow$ UAG). Consequently, the "amber" codon is precisely UAG, and this codon can perform a terminating function in mRNA [BRENNER et al., 1965; WEIGERT, GAREN, 1965 (2)].

Another group of "nonsense" mutations is known, the so-called "ochre" mutations, essentially analogous to the "amber" mutations, with the difference that the "ochre" mutations are not suppressed in the strains of bacteria that suppress the "amber" mutations (BRENNER, BECKWITH, 1965). This circumstance means that in this case interruption of growth of the polypeptide chain is determined by some other codon. It was concluded that the "ochre" mutation consists of a change of the codon of glutamine CAA into the triplet UAA (BRENNER et al., 1965).

Finally, not long ago the mutants anologous to the "amber" or "ochre" mutants were isolated, but the termination of the polypeptide chain growth was induced in them by the codon UGA (BRENNER et al., 1967; SAMBROOK et al., 1967).

Thus, three codons are found which apparently determine the interruption or termination of translation — UAG ("amber"), UAA ("ochre") and UGA.

However, it is not known whether all the three indicated triplets are actually used in the normal cell as codons signifying the end of the cistron, or whether Nature uses predominantly or exclusively only one of them. It is known that among the mutants suppressing "ochre" mutations, only those that possess a comparatively low efficiency of suppression are viable, while in mutants that suppress the "amber" mutations, and mutations with the appearance of the UGA codon, the efficiency of suppression may be very high. This seems to be indirect evidence that only the codon UAA, and not UAG or UGA, is most likely of biological significance in the normal termination of translation at the end of each cistron in the cell (BRENNER et al., 1965; SAMBROOK et al., 1967).

b) On the Mechanism of Termination

Hence, evidently a definite combination of nucleotides in the chain of the template polynucleotide serves as a signal for termination of translation and release of the polypeptide chain formed from its bond with tRNA and from the ribosome.

The cessation of translation itself, when a nonsense triplet of nucleotides is entering the ribosome, seems quite understandable and does not require any special hypothetical mechanism for its explanation. Actually, if a given triplet does not code any of the amino acids, then the C-end of the polypeptide does not form the next peptide bond, and the working cycle of the ribosome is thereupon interrupted.

However, all the complexity lies in the second aspect of the termination stage: immediately after translation has stopped, or, simultaneously and synchronously with its stopping, there should be a release of the polypeptide from the bond with tRNA and from the ribosome. It has already been indicated that the cessation of translation itself does not automatically cause release either of the polypeptide or of peptidyl-tRNA. (For example, translation evidently stops when the reading of the chain of the synthetic template has ended at its 3′-end, but there is no automatic release of peptides or peptidyl-tRNA in this case.) From this it is quite evident that the termination codon must not simply stop the translation, but must in some special way induce the release of the polypeptide.

Two principal ways of the releasing action of the termination codon are possible: 1. it induces release of peptidyl-tRNA from the ribosome, and then outside the ribosome, a special enzyme may hydrolyze the ester bond between the peptide and tRNA;

2. it somehow induces precisely hydrolysis of the ester bond between the peptide and tRNA, which automatically leads to release of the peptide from the ribosome. In all probability the second way is realized (GANOZA, NAKAMOTO, 1966). In any case, no enzyme that might specifically hydrolyze the ester $-C(=O)-O-CH<$ bond in the free peptidyl-tRNA without affecting the integrity of the peptide and tRNA themselves has been detected in homogenates of *E. coli* (GANOZA, NAKAMOTO, 1966). On the other hand, peptidyl-tRNA's have never been detected outside the ribosomes during the functioning of a cell-free system with natural templates. If, however, peptidyl-tRNA is for some reason released from the ribosome, it prevents the formation of free peptide in the system (GANOZA, NAKAMOTO, 1966). Consequently, most likely the hydrolytic release of the peptide from the bond with tRNA *occurs on the ribosome* and is a primary event, while its release from the ribosome is a consequence of this, i.e., a secondary event.

It is not yet clear through what mechanism the termination codon induces hydrolysis of the ester bond between peptidyl and the tRNA residue in the ribosome. The hypothesis exists that there is a special type of RNA in the cell, analogous to tRNA, which recognizes the termination codon with its anticodon (BRENNER et al., 1965). According to this hypothesis, the indicated "terminator" RNA, in the presence of the termination codon, specifically enters the ribosome (in its tRNA-binding site), and precisely it, and not the termination codon itself, induces the release of the peptide from its bond with the last tRNA.

Several alternative mechanisms for the releasing action of the hypothetical "terminator" RNA could be proposed. 1. "Terminator" RNA is analogous to aminoacyl-tRNA, but at the 3′-end it carries not the usual aminoacyl residue, but some special group; in the course of the working cycle of the ribosome, the C-end of the peptide is transferred to this group, but the bond formed is labile and is spontaneously hydrolyzed, releasing the polypeptide. 2. "Terminator" RNA carries an esterase component covalently bound to it, and this esterase component simply hydrolyzes the ester bond between peptidyl and tRNA when it enters the ribosome and is properly oriented next to peptidyl-tRNA. 3. "Terminator" RNA is an activator of the esterase activity of one of the ribosomal proteins, which is localized on the peptidyl-tRNA-binding site and neighbors the ester $\left(-C(=O)-O-CH<\right)$ group of peptidyl-tRNA; the entering of the "terminator" RNA activates this ribosome-bound esterase, which leads to the release of the peptide. 4. The binding of the "terminator" RNA with the ribosome induces the adding to the ribosome of a special soluble protein factor, which in combination with the ribosome brings about the hydrolysis of the ester bond of the peptidyl-tRNA.

The discovery of a special soluble protein factor necessary for the release of polypeptides (CAPECCHI, 1967) represents an argument in favor of the latter of the mentioned probabilities. However, yet another possibility should not be excluded: the protein "releasing factor" may directly recognize the termination codon and hydrolize the ester bond of the peptidyl.

Of course, all the enumerated possible mechanisms can be considered as no more than preliminary, purely speculative, hypothetical schemes, which may prove to some degree useful for the formulation of further experimental work.

c) Deformylation of the N-End of the Polypeptide

In some cases the release of the polypeptide from the ribosome may be accompanied by certain modifying reactions, completing the formation of the protein molecule from the synthesized polypeptide, although not participating directly in the termination of translation.

As has already been analyzed in detail in the section on the initiation of translation, the first N-terminal amino acid residue in the synthesis of polypeptide chains of proteins in cell-free bacterial systems *(E. coli)* is the N-formylmethionine residue [Adams, Capecchi, 1966; Webster et al., 1966; Capecchi, 1966 (1)]. At the same time, the completed proteins synthesized in bacteria *(E. coli)*, as a rule, posses a free, nonformylated, N-end (Waller, 1963). Moreover, although methionine is frequenly the N-terminal amino acid in completed proteins of *E. coli*, this is far from always being so. Consequently, to ensure the formation of entirely completed protein chains in bacterial systems, some supplementary modification of the N-end of the newly synthesized polypeptide must be accomplished; the modification, as might be thought, consists of splitting off of the terminal formyl or formylmethionyl group [Capecchi, 1966 (1)].

At what stage of protein synthesis does this modification occur, i.e., at what stage is formyl or formylmethionyl split off from the N-end of the polypeptide chain? An investigation of the kinetics of the splitting off of labeled formyl indicates that this splitting off occurs not during translation, but more likely after the completion of synthesis of the entire polypeptide chain [Capecchi, 1966 (1)]. It is believed that the splitting off of the formyl or formylmethionyl residue from the N-end of the newly synthesized polypeptide chain of the protein is accomplished with the aid of a special enzyme system. The enzyme (or enzymes) is evidently present in crude extracts of *E. coli* cells [Capecchi, 1966 (1)]. It may be that the enzyme is not present in more or less purified cell-free systems, and in any case, evidently is not associated with the ribosome.

It is important to note that evidently only the formyl residue is split off in the synthesis of some proteins of *E. coli*, and then the N-terminal amino acid is found to be methionine. However, the formation of other proteins in the same cells occurs with splitting off of formylmethionine, as a result of which other amino acids are found at the N-end, alanine, serine, threonine, etc. On the whole, the following ratio of the N-terminal amino acids is observed in completed total protein of *E. coli*: methionine — 45%, alanine — 30%, serine — 15%, threonine, glutamic acid, and other amino acids — 10% (Waller, 1963).

The substantial proportion of N-terminal alanine and serine evidently is an indication that sequences N-formylmethionyl-alanyl- and N-formylmethionyl-seryl- are very frequently encountered in polypeptide chains synthesized in *E. coli.* But it is interesting that a substantial proportion of the N-terminal methionine originates only from the formylmethionine itself and not from N-terminal sequences of the type N-formylmethionyl-methionyl-..., which are practically not detected in the synthesized polypeptide chains [CAPECCHI, 1966 (1)]. What determines the splitting off only of the formyl residue in some cases, and of the entire formylmethionyl residue from the newly formed polypeptide chain of the protein in other cases is still unknown. It can only be assumed that the deciding factor controlling the reaction is the three-dimensional conformation of the protein [CAPECCHI, 1966 (1)].

4. On the Intercistronic Punctuation in Polycistronic mRNA's

If each cistron of mRNA, i.e., the functional region coding the complete polypeptide chain of a given protein, begins with an initiation codon and ends with a termination codon, then for the case of polycistronic mRNA's two models of reading out of the cistrons by the ribosome can be considered. 1. Independent attachment of the ribosomes to the initiation codon of each cistron, reading out, and then, immediately after termination, detachment of the ribosome from the mRNA chain

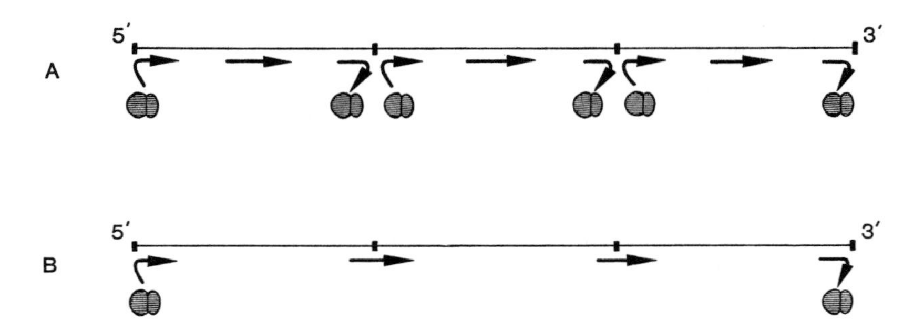

Fig. 21. Two models of reading out of polycistronic mRNA by ribosomes: A Independent reading out of cistrons; B Successive reading out of cistrons

(Fig. 21, A). In this case, the cistrons are read out quite independently of one another. 2. Attachment of the ribosomes only to the 5'-terminal initiating codon, reading out of the first cistron, then termination without detachment of the ribosome from mRNA, arrival of the ribosome at the initiation codon of the following cistron, reading out of the following cistron, etc. This means a successive dependent reading out of the cistrons, from the 5'-end of the chain to the 3'-end (Fig. 21, B). If the first model is correct, then any interruption in the reading out or shift of the phase of the reading out in any preceding cistron should not influence the reading out of the following cistrons of the mRNA chains. Experience shows that this is not so, and that interruption of translation in the case of a "nonsense" mutation or shift of the phase of reading out in one of the preceding cistrons influences the reading out of the cistrons situated in the direction toward the 3'-end from the disturbed cistron (polar effect of mutation). Thus, the second model is confirmed, i.e., *the attachment of the ribosomes only to the 5'-end of polycistronic mRNA and reading out of the cistrons in strict sequence, one after another* (AMES, HARTMAN, 1963; YANOFSKY, ITO, 1966; MARTIN et al., 1966).

The appearance of "amber" or "ochre" mutation in one of the initial cistrons leads to a decrease in protein synthesis on the unaltered cistrons, situated in the direction toward the 3′-end of the polycistronic template, i.e., the "amber" and "ochre" mutations give a polar effect (NEWTON et al., 1965; ITO, CRAWFORD, 1965; YANOFSKY, ITO, 1966; MARTIN et al., 1966). As a rule, the closer the "amber" or "ochre" mutation to the beginning of the cistron, the greater the polar effect, i.e., the greater the degree to which protein synthesis on the following cistrons is reduced (NEWTON et al., 1965; BAUERLE, MARGOLIN, 1966; YANOFSKY, ITO, 1966). These facts are explainable if we assume that: a) termination of translation and release of the peptide from the ribosome weaken the association of the ribosome with the template; b) a non-translating ribosome may "slip" down the mRNA chain (YANOFSKY, ITO, 1966; MARTIN et al., 1966). Then, the greater the distance that the non-translating ribosome must "slip" along the mRNA chain to the initiating triplet of the following cistron, the greater the probability of dissociation of the ribosome from mRNA; hence in the case of a large distance, fewer ribosomes reach the following cistron than in the case of a small distance, and consequently, the number of synthesized protein molecule on the remaining region of the mRNA chain will be smaller (Fig. 22).

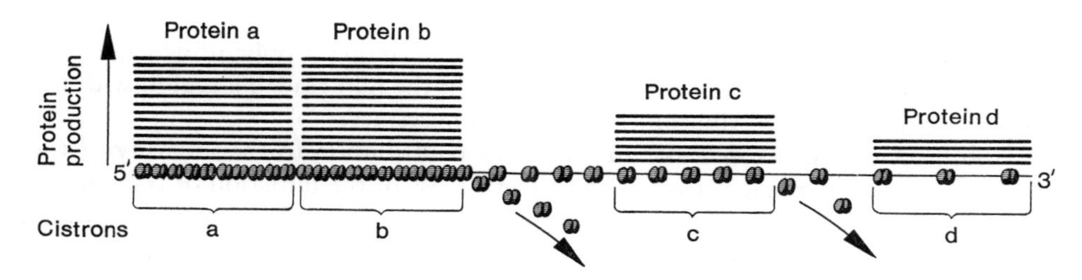

Fig. 22. "Polar effect" of non-readable intercistronic regions in the chain of polycistronic mRNA: dependence of protein production of the cistrons on the length of the preceding non-readable intercistronic regions (scheme)

In the normal termination of translation of a cistron in polycistronic mRNA, there can be two cases: 1. The initiation region of a following cistron immediately adjoins the termination codon (.....UAAAUG...), or, perhaps, even overlaps it (.....UAAUG....). In this case, the ribosome, having completed the translation of a preceding cistron, immediately begins translation of the following cistron, without being able to dissociate from mRNA. Then the indicated adjacent cistrons should always be read out with *the same intensity*, giving a strictly constant, equimolar, ratio of the proteins corresponding to them (Fig. 22). It is known that such groups of coordinately working cistrons actually exist (WILSON, HOGNESS, 1964; WILSON, CRAWFORD, 1965; CREIGHTON, YANOFSKY, 1966; MARTIN et al., 1966) 2. Between the termination codon of a preceding cistron and the initiation sequence of the following cistron, there is *a non-translatable region of mRNA of greater or smaller length* (Fig. 22). In such a case, the ribosome, after completion of translation of the preceding cistron, has to "slip" down along the non-translatable intercistronic region. In view of the definite probability of dissociation of the non-translating ribosome from mRNA, fewer ribosomes compared with those that have passed along the preceding cistron, will reach the following cistron. Hence, in the case of such relations between the cistrons, a difference should be observed in the intensity of the synthesis of various proteins coded by polycistronic mRNA, i.e., *a cistron situated farther from the 5'-end of mRNA should be read out less intensively than the preceding cistron* (Fig. 22). Such relationships are also observed in actual situations (ZABIN, 1963; AMES, HARTMAN, 1963).

Certain data, however, are apparently an indication that the reduced effectiveness of the reading out of a following cistron may be a function *not only of the length of the non-readable intercistronic region* (MARTIN et al., 1966). Thus, the presence of a non-readable region between the termination codon and the next initiation codon does not always appreciably reduce the effectiveness of the reading out of the next cistron. On the other hand, there are examples when in the case of a short distance between the termination and initiation codons, a great decrease in the effectiveness of reading out of the next cistron is observed (strong polarity).

Perhaps a great role in the dissociation of ribosomes from mRNA on an intercistronic region may be played not only (and perhaps not so much) by its length, but also by its nucleotide sequence (the presence of some sort of special nucleotide combinations) and secondary structure. For example, intensive dissociation of the non-reading ribosomes from mRNA could occur when a helical region is encountered. The non-reading ribosomes may be relatively weakly retained on the polypurine sequences. Therefore, a full direct correlation of the length of the non-readable region with the proportion of dissociated ribosomes is not at all necessary.

Another factor that can make an important contribution to the intensity of the reading out of an internal cistron is the effectiveness of the initiation of translation, when the ribosome reaches its initiation codon. The matter is that the presence of an initiation triplet itself, for example, AUG, in some *internal* region of the mRNA chain may be insufficient for the initiation of translation by a "slipping" non-translating ribosome. Actually, in the case of "amber" and "ochre" mutations, translation is interrupted at a corresponding point of a cistron, and the ribosome then "slips" down along the remainder of the cistron without translation. In this case, it should encounter methionine codons AUG and valine codons GUG, each of which might be an initiation codon; nonetheless, translation for some reason is not initiated, i.e., C-terminal peptide fragments of proteins are not synthesized. Moreover, as has been shown, the "slipping" of the non-translating ribosome along the mRNA chain evidently occurs with no defined phase (MARTIN et al., 1966; SARABHAI, BRENNER, 1967); then the number of AUG and GUG combinations encountered along the way is actually even greater, but again there is no initiation. From this it should be assumed that for the initiation of translation by an internal initiation codon, the neighborhood of some special nucleotide combinations is required; these nucleotide combinations must most likely directly precede the initiation codon; their function must consist of making the non-translating ribosome competent for initiation. The effectiveness of hypothetical preinitiating combinations of nucleotides in the conversion of the ribosome to the competent state may be varied, and if it is not great, then only a portion of the ribosomes, from those that have reached the beginning of a given cistron, will be capable of initiating translation. Thus, the effectiveness of the initiation of translation of a following cistron also can determine the degree of reduction of the intensity of its reading out in comparison with the preceding cistron, i.e., make its own contribution to the degree of polarity. The hypothesis that it is the effectiveness of the initiating region of a following cistron, and not intrinsically the length of the intercistronic non-translatable region that can be the main factor determining polarity, was recently advanced by the group of MARTIN and AMES (MARTIN et al., 1966).

In any case, independent of the correctness of one or another hypothesis explaining polarity, it should be emphasized that a number of considerations and indirect data indicate the need not only for the termination codon at the end, and the initiation codon at the beginning of each cistron, but also for special *preinitiating nucleotide combinations at the boundary between the cistrons* in polycistronic mRNA's. It is not excluded that the recently detected "starter" mutations leading to the reinitiation of translation after its termination by the "amber" or "ochre" codon (SARABHAI, BRENNER, 1967), demonstrate the appearance of pre-initiating nucleotide sequence (the "starter") after which, in phase, the translation may be initiated immediately.

IV. Appendix: On the Mechanism of the Action of Certain Antibiotics

The bactericidal and bacteriostatic effects of a number of well known antibiotics — streptomycin, chloramphenicol, tetracyclines, puromycin, and certain others — have proved to be directly due to the fact that they affect bacterial ribosomes and disturb their normal functioning.

1. Puromycin

The mechanism of the action of this antibiotic has been best studied. In chemical nature it represents a nucleoside derivate — 6-dimethylamino-9-(3′-*p*-methoxy-L-phenylalanylamino-3′-deoxy-β-D-ribofuranosyl) purine (Fig. 23).

Even a simple consideration of the chemical formula shows that puromycin is a structural analog of the 3′-terminal aminoacylated group of tRNA (YARMOLINSKY, HABA, 1959). Hence, the inhibiting effect of puromycin upon protein synthesis may be explained by competitive replacement of following aminoacyl-tRNA during the process of translation. Direct experimental data indicate that this is so and that the result of its addition to a translating ribosome is the formation of a peptide bond between it and the growing peptide, with the release of the peptidyl-puromycin from the ribosome (GILBERT, 1963; NATHANS, 1964; TRAUT, MONRO, 1964; SMITH et al., 1965)[1].

Thus, puromycin evidently participates in the step of formation of a peptide bond in the ribosome, replacing the following aminoacyl-tRNA: transfer of the C-end of the growing peptide chain from peptidyl-tRNA occurs not to the aminoacyl-tRNA, but to the puromycin replacing it — to the free amino group of its aminoacyl residue:

$$\text{peptidyl-tRNA} + \text{puromycin} \rightarrow \text{peptidyl-puromycin} + \text{tRNA}.$$

The peptidyl-puromycin formed cannot be retained by the ribosome and is released from it into the medium. The ribosome remains "empty," i.e., without a growing peptide. It is this that means a break off of translation.

[1] The release of the nascent peptide from the ribosome under the action of puromycin and the fact that puromycin is combined with the chain of the released peptide were first demonstrated on animal systems (MORRIS, SCHWEET, 1961; ALLEN, ZAMECNIK, 1962).

A knowledge of the fact that puromycin does not inhibit, in the real sense of the word, any of the enzymatic stages of protein synthesis, but itself participates in the formation of a peptide bond in the ribosome permits its use for the experimental study of certain aspects of the functioning of the ribosome.

Thus, it has been shown that the reaction of puromycin requires neither GTP nor protein transfer factors [Traut, Monro, 1964; Rychlik, 1966 (1); Bretscher, Marcker, 1966; Zamir et al., 1966; Monro, Marcker, 1967; Monro, 1967]. Consequently, the act of formation of a peptide bond in the ribosome occurs without the participation of GTP and without any of the transfer factors, but *is evidently catalyzed by the ribosome itself*. From this it is clear that not only the G-factor, but also both T-factors have no relationship to this act of the working cycle and can in no way be considered as the "polymerizing enzyme," "peptide synthetase," etc.; rather, the point

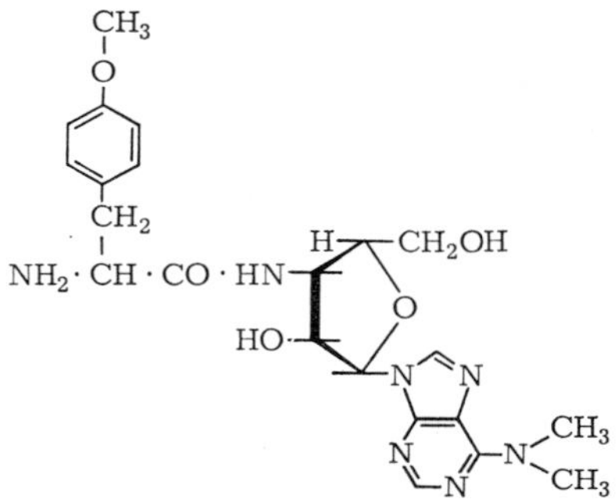

Fig. 23. Puromycin

of application of all of them together should be sought in the other steps of the working cycle. A confirmation of this is the fact that *para*chlormercuribenzoate, which inactivates transfer factors, when added to a protein-synthesizing system, just as one should have expected, entirely inhibits translation as a whole but does not at all inhibit the puromycin reaction and the release of peptides from the ribosomes (Traut, Monro, 1964).

At the same time, the integrity of the entire ribosome is not necessary for the reaction with puromycin: a peptide bond between the peptide and puromycin is also formed if isolated 50 S subparticles, which retain peptidyl-tRNA, are present in the system (Traut, Monro, 1964). Consequently, *the peptidyl-transferase ("peptide-synthetase") catalytical center of the ribosome is localized on the 50 S subparticle.*

It is known that when puromycin is added to the translating ribosome, only peptidyl-puromycin can be formed, but not aminoacyl-puromycin (Smith et al., 1965). Consequently, peptidyl-tRNA reacts with puromycin, while aminoacyl-tRNA does not react with it. This is most likely an indication that peptidyl-tRNA and aminoacyl-tRNA are localized on different sites of the ribosome — in the peptidyl-tRNA-binding site of the 50 S subparticle and in the aminoacyl-tRNA-binding site of the 30 S subparticle respectively. Puromycin can attack the corresponding ester group only when the group is set up on the 50 S subparticle.

However, although all the usual aminoacyl-tRNA's in the ribosome do not react with puromycin, formylmethionyl-tRNA or methionyl-tRNA$_{f\text{-met}}$, bound by the ribosome in the presence of the initiation codon, reacts with puromycin just as well as does peptidyl-tRNA (Bretscher, Marcker, 1966; Zamir et al., 1966):

$$\text{formylmethionyl-tRNA}_{\text{f-met}} + \text{puromycin} \rightarrow \text{formylmethionyl-puromycin} + \text{tRNA}_{\text{f-met}}.$$

Consequently, the initiator tRNA actually occupies not the aminoacyl-tRNA-binding site on the ribosome, designed for the usual aminoacyl-tRNA's, but the peptidyl-tRNA-binding site on the 50 S subparticle.

It has been noted that under the action of puromycin on working ribosomes, in the complete system of protein synthesis, up to 100% of the peptides are released. However, if translation is stopped, for example, by removing the energy sources and transfer factors, puromycin releases only part, about half, of the nascent peptides when it is used to treat ribosomes (Traut, Monro, 1964; Cannon, 1967). This may be explained on the basis of the working cycle of the ribosome (see Section III, 2, e). If the work of a ribosome is stopped at the moment when the tRNA residue of the peptidyl-tRNA molecule is in its peptidyl-tRNA-binding site on the 50 S subparticle (Fig. 20, top left), then the reaction with puromycin, i.e., transfer of the C-end of the peptide to it, is permitted. If, however, the ribosome is stopped at the moment when the C-end of the peptidyl residue has just been transferred to the aminoacyl-tRNA in the aminoacyl-tRNA-binding site, and translocation has not yet occurred (Fig. 20, bottom left), then the reaction with puromycin is evidently impossible. The reaction with puromycin is also evidently hindered in the case when the place next to the C-end of the peptide is occupied by the aminoacylated adenosine end of aminoacyl-tRNA (Fig. 20, bottom right). In any case the fact that only part of the non-functioning ribosomes releases peptides under the action of puromycin immediately permitted the conclusion to be drawn that two different states of the ribosome are possible during its functioning (Traut, Monro, 1964).

Hence, puromycin acts as a competitive analog of aminoacyl-tRNA, replacing the latter in the reaction with peptidyl-tRNA. The reaction of puromycin with peptidyl-tRNA leads to the release of the peptidyl from the ribosome in the form of peptidyl-puromycin, i.e., to a cessation of protein synthesis. Puromycin reacts with peptidyl-tRNA only in the ribosome. Puromycin does not react with the usual aminoacyl-tRNA's bound to the ribosome, but does react with bound initiator aminoacyl-tRNA–formylmethionyl-tRNA. The formation of a peptide bond between puromycin and the bound peptide residue requires neither GTP nor any transfer factors: the reaction is catalyzed by the ribosome itself. The presence of only the 50 S subparticles, retaining the peptidyl-tRNA, is sufficient to bring about the reaction with puromycin. However, when puromycin is used to treat ribosomes in which translation has been stopped by removing the transfer factors or energy sources, only part of the peptidyl-tRNA reacts with puromycin, whereas the other part does not; this may indicate that in such a stopped system, the ribosomes are in two different states.

2. Chloramphenicol (Chloromycetin)

It has been shown that this antibiotic also blocks protein synthesis at the ribosome level (Nathans, Lipmann, 1961; Nathans et al., 1962; Rendi, Ochoa, 1962) However, in this case, no conclusions can yet be drawn on the possible point of application of the action of chloramphenicol upon the ribosomal protein synthesis from a consideration of the structural formula (Fig. 24). In contrast to puromycin, which acts both upon bacterial and animal systems, chloramphenicol is more specific, and as a rule, exerts a strong inhibiting effect primarily upon bacterial ribosomes. Data have been obtained indicating that it can be very loosely bound to the bacterial ribosome [VAZQUEZ, 1963, 1964, 1966 (1); WOLFE, HAHN, 1965; DAS et al., 1966]. The site for this labile binding is localized on the 50 S subparticle. It has been shown that chloramphenicol influences neither the ability of the ribosome to bind the template polynucleotide nor the specific binding of aminoacyl-tRNA and the formation of the ternary complex of ribosome–template–aminoacyl-tRNA, nor the formation of the initial complex (SPEYER et al., 1963; KUČAN, LIPMANN, 1964; NAKAMOTO et al., 1963; WOLFE, HAHN, 1965; DAS et al., 1966; WEBER, DE MOSS, 1966). Neither does it inhibit the binding of peptidyl-tRNA to the peptidyl-tRNA-binding site on

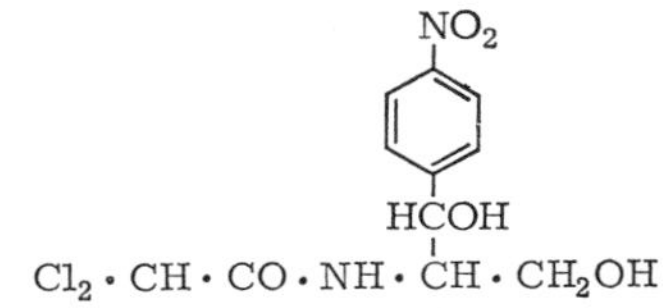

NO_2

HCOH

$Cl_2 \cdot CH \cdot CO \cdot NH \cdot CH \cdot CH_2OH$

Fig. 24. Chloramphenicol (chloromycetin)

the 50 S subparticle of the ribosome [RYCHLIK, 1966 (1, 2)]. At the same time, it entirely inhibits the reaction of puromycin with peptidyl-tRNA bound on the ribosome [NATHANS et al., 1962; NATHANS, 1964; TRAUT, MONRO, 1964; RYCHLIK, 1966 (1)]. Acting upon cells and cell-free protein-synthesizing systems, chloramphenicol stops the growth of the peptide in the ribosome; the peptide is not released in this case, but remains bound within the ribosome (JULIAN, 1965; DAS et al., 1966; WEBER, DE MOSS, 1966). From all the above mentioned, it can be already assumed that chloramphenicol directly blocks the reaction of the peptide bond formation in the ribosome due to inhibition of the peptide-synthetase activity of the ribosomal particle itself. Recently it has been proved by more direct experiments that the point of application of the inhibitor is the peptidyl-transferase (peptide synthetase) center on the 50 S subparticle (MONRO, 1967; MONRO, VAZQUEZ, 1967).

Probably the antibiotic *erythromycin* acts similarly [VAZQUEZ, 1963, 1966 (1, 2); TAUBMAN et al., 1963, 1966; WOLFE, HAHN, 1964, 1965; TANAKA, TERAOKA, 1966; RYCHLIK, 1966 (1); VAZQUEZ, MONRO, 1967]. However, in contrast to chloramphenicol, it is rather firmly bound to the 50 S subparticles of the ribosomes from erythromycin-sensitive bacteria. Ribosomes from resistant mutants bind the antibiotic far more weakly (TAUBMAN et al., 1966).

3. Tetracyclines

Tetracycline antibiotics represent still another group of substances that specifically inhibit protein synthesis at the ribosomal level (RENDI, OCHOA, 1962; FRANKLIN, 1963). The structural formulas of the most frequently used antibiotics of this group, tetracycline and chlortetracycline, are given in Fig. 25. There is an indication that the tetracyclines are selectively bound to the 30 S subparticle of the ribosome (Connamacher, Mandel, 1965). It has been shown that tetracyclines in concentrations inhibiting protein synthesis (10^{-5} to 10^{-4} M) inhibit precisely the specific binding of aminoacyl-tRNA to the ribosome in the presence of the template (Suarez, Nathans, 1965; Hierowski, 1965; Vazquez, Monro, 1967). In exactly the same way, the specific binding of aminoacyl-tRNA to the isolated 30 S subparticle in the presence of template polynucleotide is inhibited by tetracycline (Suzuka et al., 1966; Vazquez, Monro, 1967). In this case, the ability of the ribosome to associate with the template polynucleotide itself is retained. Tetracyclines in a concentration of 10^{-4} M do not inhibit the attachment of peptidyl-tRNA to the ribosome [Rychlik, 1966 (2)]. Consequently, it might be thought that tetracyclines act selectively only *upon the aminoacyl-tRNA-binding site of the 30 S subparticle*, most likely sticking directly to the site. However, if aminoacyl-tRNA is already bound to the ribosome or to the

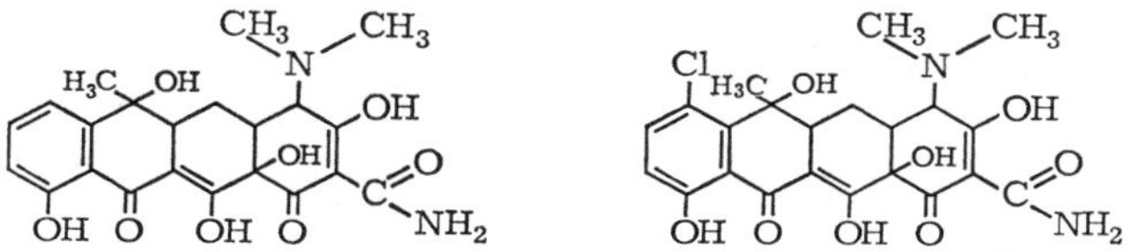

Fig. 25. Tetracycline (left) and chlortetracycline (right)

30 S subparticle, tetracycline does not displace it from the complex (Hierowski, 1965; Suzuka et al., 1966).

At the same time, it has been shown that in the presence of relatively high concentrations of tetracycline (10^{-3} M), when translation is entirely suppressed, puromycin releases only about half of the peptides from tRNA and ribosomes (Traut, Monro, 1964). It is possible that this is just a nonspecific effect of the high antibiotic concentration upon the reaction of the peptide bond formation (Monro, Vazquez, 1967). But another possibility also cannot be entirely excluded, that at high concentrations tetracycline can act upon the peptidyl-tRNA-binding site on the 50 S subparticle as well and thereby block translocation; in this event, just as in the case of removal of energy sources (see above, "Puromycin"), only half the ribosomes of the system will be competent for the action of puromycin.

It has been found that the antibiotic *edeine*, just like tetracycline, inhibits the specific binding of aminoacyl-tRNA to the bacterial ribosomes; but in contrast to the tetracyclines, the edeine molecule evidently combines simultaneously with both subparticles, the 50 S and 30 S, and even in the presence of a low concentration of Mg^{++} causes the association of the 30 S and 50 S subparticles into the 70 S ribosome (Hierowski, Kurylo-Borowska, 1965; Kurylo-Borowska, Hierowski, 1965).

4. Streptomycin and Other Aminoglucoside Antibiotics

Streptomycin, a carbohydrate substance with strongly basic properties (Fig. 26), is one of the most widely used antibacterial antibiotics. It has been shown that the antibiotic acts upon thep rotein-synthesizing apparatus of strains of bacteria sensitive to it and that this action is due to the sensitivity of their ribosomal apparatus to the antibiotic (Erdös, Ullmann, 1959, 1960; Hancock, 1961; Spotts, Stanier, 1961). Correspondingly, it inhibits a cell-free protein-synthesizing system containing ribosomes from streptomycin-sensitive strains and does not inhibit a system in which the ribosomes were obtained from streptomycin-resistant cells [Flaks et al., 1962 (1, 2); Speyer et al., 1962; Mager et al., 1962]. In addition to inhibition, the antibiotic causes errors in translation, or misreading of the template polynucleotide, i.e., it relaxes the strict correspondence between the codons and amino acids incorporated into the polypeptide, and once again this is observed only with ribosomes from streptomycin-sensitive bacteria and is not observed with ribosomes from streptomycin-resistant strains (Davies et al., 1964). The site of application of the strepto-

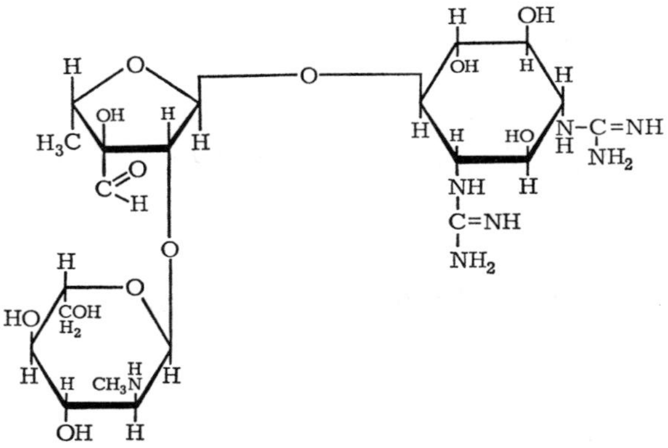

Fig. 26. Streptomycin

mycin action upon the ribosome is its 30 S subparticle (Cox et al., 1964; Davies, 1964). Probably streptomycin interacts specifically with some definite structural protein of the 30 S subparticles of the ribosomes sensitive to streptomycin; a mutation resulting in the acquisition of resistance to streptomycin may consist of a change in this protein, leading to a loss of the affinity for streptomycin.

The interaction of streptomycin with the 30 S subparticle of the ribosome does not change its ability to bind the template polynucleotide (Davies, 1964; Kaji, Kaji, 1965; Kaji, 1966). All the available data indicate that the antibiotic acts upon the specific binding of the aminoacyl-tRNA to the ribosome in the presence of the template (Kaji, Kaji, 1965; Pestka et al., 1965; Pestka, 1966). However, this action is complex, and in all probability cannot be reduced to a direct blocking of the aminoacyl-tRNA-binding site of the 30 S subparticle.

If isolated 30 S subparticles are investigated, it is found that streptomycin, as a rule, partially inhibits the binding of aminoacyl-tRNA in a ternary complex with the 30 S subparticle and template polynucleotide (Kaji et al., 1966; Kaji, 1966). The effect is very specific, and in the case of 30 S subparticles of streptomycin-resistant strains, no inhibition of binding is observed. The inhibiting effect of streptomycin upon the specific binding of aminoacyl-tRNA is also manifested in a system with complete (70 S) ribosomes, taken from streptomycin-sensitive cells (Kaji, Kaji, 1965; Pestka, 1966).

However, on complete ribosomes a different, stronger effect of streptomycin may also be observed: there is a stimulation of the binding of the aminoacyl-tRNA's that do not correspond to the codons of the template polynucleotide. For example, ribosomes with polyU, which normally bind phenylalanyl-tRNA (UUU codes phenylalanine — Fig. 5), and only very weakly bind isoleucyl-tRNA, begin to bind isoleucyl-tRNA considerably more actively in the presence of streptomycin (isoleucine is coded, in particular, by the triplet AUU — Fig. 5); the binding of leucyl-tRNA (coded, in particular, by the triplets UUA, UUG, CUU) and seryl-tRNA (coded by the triplets UCU, UCC, UCA, and UCG) is also stimulated by streptomycin (KAJI, KAJI, 1965; PESTKA et al., 1965). Evidently this effect is basically responsible for the miscoding (misreading) observed in cell-free protein-synthesizing systems in the presence of streptomycin; in particular, it was found that in a system with polyU, the synthesized polypeptides contain not only phenylalanine, but also isoleucine, leucine, and serine (DAVIES et al., 1964, 1965). This effect is also very specific and is not observed with ribosomes from streptomycin-resistant strains. The effect of misbinding of aminoacyl-tRNA in the presence of streptomycin is manifested *only on complete ribosomes*; in systems with isolated 30 S subparticles + template it is absent (PESTKA, NIRENBERG, 1966; KAJI, 1966). Misbinding of aminoacyl-tRNA, caused by streptomycin in systems of 70 S ribosomes + template, can be substantially reduced or even practically eliminated if the ribosomes used in the system were preliminarily dissociated into their subparticles and then reassociated (PESTKA, 1966; KAJI, 1966). (It is known that the association between the subparticles in reassociated 70 S ribosomes is for some reason appreciably weaker than in non-dissociated 70 S ribosomes.) It might be thought, consequently, that the effect of misbinding of aminoacyl-tRNA's is due not so much to the 30 S subparticle itself as to the interaction between the 30 S and 50 S subparticles or the presence of some sort of component, labilely retained between the subparticles in the ribosome.

Although the available data are rather complex and confusing, they can still be used as the basis for preliminary conclusions on the action of streptomycin upon ribosomes sensitive to it. 1. Streptomycin selectively interacts with the 30 S subparticle of the ribosome. 2. This interaction does not occur at the mRNA-binding center nor at the aminoacyl-tRNA-binding site of the 30 S subparticle. 3. The interaction of streptomycin with the 30 S subparticle leads to two consequences: a) to a certain decrease in the affinity of the aminoacyl-tRNA-binding site of the 30 S subparticle for aminoacyl-tRNA; a consequence of this is a partial inhibiting effect upon the specific binding of aminoacyl-tRNA; b) to changes at the boundary between the subparticles, possibly to the involving of some groups of the 50 S subparticle or of a component at the boundary between the subparticles in the process of binding of aminoacyl-tRNA (codon–anticodon interaction); a consequence of this is the effect of misbinding of aminoacyl-tRNA's. Weakening of the contact between the subparticles or complete dissociation eliminates the misbinding effect, leaving the inhibiting effect untouched. 4. Both effects of streptomycin are indirect in the sense that the functional groups of streptomycin themselves do not participate directly either in the retention of the "mistaken" aminoacyl-tRNA on the ribosomal particle or in the repulsion of the "correct" aminoacyl-tRNA. Most likely both effects are a consequence of the distorting influence of streptomycin upon the conformation and distribution of charges in the 30 S subparticle as a whole. 5. A general result of the effect of streptomycin upon the process of translation is its ultimate intervention into the process of specific entering of aminoacyl-tRNA into the ribosome; in this case it is more the misreading of a template that takes place than the inhibition of translation.

A whole series of other antibiotics of the aminoglucoside type are now known (kanamycin, neomycin, etc.), which act upon the bacterial synthesis of protein similarly to streptomycin, causing miscoding (Davies et al., 1964, 1965, 1966). They all are, just like streptomycin, strongly basic substances, containing amine or guanidine groups. Kanamycin and neomycin evidently act more strongly than streptomycin in a number of cases, stimulating the erroneous incorporation of an even larger set of amino acids.

References

Preface. Fundamental Introduction

Belozersky, A. N., and A. S. Spirin: Nature **182**, 111 (1958).
— — The Nucleic Acids, **3** p. 147. New York: Acad. Press 1960.
Bessman, M. J., I. R. Lehmann, E. S. Simms, and A. Kornberg: J. Biol. Chem. **233**, 171 (1958).
Bollum, F. J.: Progress in Nucleic Acid Research, **1**, p. 1. New York: Acad. Press 1963.
Brenner, S., F. Jacob, and M. Meselson: Nature **190**, 576 (1961).
—, A. O. W. Stretton, and S. Kaplan: Nature **206**, 994 (1965).
Caspersson, T.: Naturwissenschaften **28**, 33 (1941).
Chargaff, E.: Experientia **6**, 201 (1950).
— J. Cellular Comp. Physiol. **38**, Suppl. 1, 41 (1951).
Crick, F. H. C.: Symp. Soc. Exptl. Biol. **12**, 138 (1958).
— Progress in Nucleic Acid Research **1**, 163 (1963).
— J. Mol. Biol. **19**, 548 (1966).
—, L. Barnett, S. Brenner, and R. J. Watts-Tobin: Nature **192**, 1227 (1961).
—, J. S. Griffith, and L. E. Orgel: Proc. Nat. Acad. Sci. US **43**, 416 (1957).
Franklin, R. E., and R. G. Gosling: Nature **171**, 737 (1953).
Gamow, G.: Nature **173**, 318 (1954).
—, A. Rich, and M. Ycas: Advances in Biol. and Med. Phys. **4**, 23 (1956).
Geiduschek, E. P., T. Nakamoto, and S. B. Weiss: Proc. Nat. Acad. Sci. US **47**, 1405 (1961).
Gros, F., H. Hiatt, W. Gilbert, C. G. Kurland, R. W. Risebrough, and J. D. Watson: (1) Nature **190**, 581 (1961).
—, W. Gilbert, H. H. Hiatt, G. Attardi, P. F. Spahr, and J. D. Watson: (2) Cold Spring Harbor Symposia Quant. Biol. **26**, 111 (1961).
Hoagland, M. B.: The Nucleic Acids, **3**, p. 349. New York: Acad. Press 1960.
—, M. L. Stephenson, J. F. Scott, L. I. Hecht, and P. C. Zamecnik: J. Biol. Chem. **231**, 241 (1958).
—, P. C. Zamecnik, and M. L. Stephenson: Biochim. et Biophys. Acta **24**, 215 (1957).
Hurwitz, J., and J. T. August: Progress in Nucleic Acid Research, **1**, p. 59. New York: Acad. Press 1963.
—, A. Bresler, and R. Diringer: Biochem. Biophys. Res. Communs. **3**, 15 (1960).
—, J. J. Furth, M. Anders, and A. H. Evans: J. Biol. Chem. **237**, 3752 (1962).
— — —, P. J. Ortiz, and J. T. August: Cold Spring Harbor Symposia Quant. Biol. **26**, 91 (1961).
Jacob, F., and J. Monod: J. Mol. Biol. **3**, 318 (1961).
Lehmann, I. R., S. B. Zimmermann, J. Adler, M. J. Bessman, E. S. Simms, and A. Kornberg: Proc. Nat. Acad. Sci. US **44**, 1191 (1958).
Marmur, J., C. Greenspan, F. Palecek, F. M. Kahan, J. Levine, and M. Mandel: Cold Spring Harbor Symposia Quant. Biol. **28**, 191 (1963).
Meselson, M., and F. W. Stahl: Proc. Nat. Acad. Sci. US **44**, 671 (1958).
Morgan, A. R., R. D. Wells, and H. G. Khorana: Proc. Nat. Acad. Sci. US **56**, 1899 (1966).
Nirenberg, M. W., and J. H. Matthaei: Proc. Nat. Acad. Sci. US **47**, 1588 (1961).
—, M. Bernfield, R. Brimacombe, J. Trupin, F. Rottman, and C. O'Neal: Proc. Nat. Acad. Sci. US **53**, 1161 (1965).

NIRENBERG, M. W., O. W. JONES, P. LEDER, B. F. C. CLARK, W. S. SLY, and S. PESTKA: Cold Spring Harbor Symposia Quant. Biol. **28**, 549 (1963).
NOLL, H.: Proc. 19th Symp. on Fundamental Cancer Research (Information Exchange Group No. 7, Scientific Memo N 99). 1965.

OGATA, K., H. NOHARA, and T. MORITA: Biochim. et Biophys. Acta **26**, 657 (1957).

PETERMANN, M. L.: The physical and chemical properties of ribosomes. Amsterdam-London-New York: Elsevier Publ. Co. 1964.

RICH, A.: Scientific American, p. 44. December 1963.
—, J. R. WARNER, and H. M. GOODMAN: Cold Spring Harbor Symposia Quant. Biol. **28**, 269 (1963).
ROBERTS, R. B.: Studies of macromolecular biosynthesis. Washington: Carnegie Institution 1964.

SPEYER, J. F., P. LENGYEL, C. BASILIO, A. J. WAHBA, R. S. GARDNER, and S. OCHOA: Cold Spring Harbor Symposia Quant. Biol. **28**, 559 (1963).
SPIEGELMAN, S.: Cold Spring Harbor Symposia Quant. Biol. **26**, 75 (1961).
SPIRIN, A. S., A. N. BELOZERSKY, N. V. SHUGAYEVA, and B. F. VANYUSHIN: Biokhimiya **22**, 744 (1957).
STEVENS, A.: Biochem. Biophys. Res. Communs. **3**, 92 (1960).

WATSON, J. D.: Science **140**, 17 (1963).
— Bull. soc. chim. biol. **46**, 1399 (1964).
— Molecular biology of the gene. New York-Amsterdam: Benjamin 1965.
—, and F. H. C. CRICK: (1) Nature **171**, 737 (1953).
— — (2) Nature **171**, 964 (1953).
WEISS, S. B.: Proc. Nat. Acad. Sci. US **46**, 1020 (1960).
—, and T. NAKAMOTO: Proc. Nat. Acad. Sci. US **47**, 1400 (1961).
WILKINS, M. H. F., A. R. STOKES, and H. R. WILSON: Nature **171**, 737 (1953).
WITTMANN, H. G., and B. WITTMANN-LIEBOLD: Cold Spring Harbor Symposia Quant. Biol. **28**, 589 (1963).

YANOFSKY, C.: Cold Spring Harbor Symposia Quant. Biol. **28**, 581 (1963).

Part One

I. Physical Properties and Chemical Composition of the Ribosomes

ABDUL-NOUR, B., and G. C. WEBSTER: Exp. Cell Research **20**, 226 (1960).

BAYLEY, S. T., and D. J. KUSHNER: J. Mol. Biol. **9**, 654 (1964).
BELITSINA, N. V., L. P. OVCHINNIKOV, A. S. SPIRIN, YU. Z. GENDON, and V. I. TCHERNOS: Molekularnaya Biologia (U.S.S.R.) **2**, 739 (1968).
BOARDMAN, N. K., R. J. B. FRANCKI, and S. G. WILDMAN: Biochemistry **4**, 872 (1965).
— — — J. Mol. Biol. **17**, 470 (1966).
BRETTHAUER, R. K., L. MARCUS, J. CHALOUPKA, H. O. HALVORSON, and R. M. BOCK: Biochemistry **2**, 1079 (1963).
BROWN, D. D., and J. D. CASTON: Develop. Biol. **5**, 412 (1962).
BRUSKOV, V. I., and N. A. KISSELEV: J. Mol. Biol. **37**, 367 (1968).

CHAO, F.-C.: Arch. Biochem. Biophys. **70**, 426 (1957).
—, and H. K. SCHACHMAN: Arch. Biochem. Biophys. **61**, 220 (1956).
CHOI, Y. S., and C. W. CARR: J. Mol. Biol. **25**, 331 (1967).
CLARK, M. F., R. E. F. MATTHEWS, and R. K. RALF: Biochim. et Biophys. Acta **91**, 289 (1964).
COHEN, S. S., and J. LICHTENSTEIN: J. Biol. Chem. **235**, 2112 (1960).

DIBBLE, W. E.: J. Ultrastruct. Research. **11**, 363 (1964).
—, and H. M. DINTZIS: Biochim. et Biophys. Acta **37**, 152 (1960).

Dintzis, H. M., H. Borsook, and J. Vinograd: Microsomal particles and protein synthesis, p. 95. London-New York-Paris-Los Angeles: Pergamon Press 1958.

Edelman, I. S., P. O. P. Ts'o, and J. Vinograd: Biochim. et Biophys. Acta **43**, 393 (1960).
Elson, D.: Biochim. et Biophys. Acta **36**, 362 (1959).
— Biochim. et Biophys. Acta **53**, 232 (1961).

Gavrilova, L. P., D. A. Ivanov, and A. S. Spirin: (1) J. Mol. Biol. **16**, 473 (1966).
—, M. I. Lerman, and A. S. Spirin: (2) Izvest. Akad. Nauk S.S.S.R., Ser. Biol., No. 6, 826 (1966).
Gesteland, R. F.: J. Mol. Biol. **18**, 356 (1966).
Goldberg, A.: J. Mol. Biol. **15**, 663 (1966).

Hall, C. E., and H. S. Slayter: J. Mol. Biol. **1**, 329 (1959).
Hamilton, M. G., and M. L. Petermann: J. Biol. Chem. **234**, 1441 (1959).
—, L. F. Cavalieri, and M. L. Petermann: J. Biol. Chem. **237**, 1155 (1962).
Hart, R. G.: Biochim. et Biophys. Acta **60**, 629 (1962).
— Biochim. et Biophys. Acta **72**, 662 (1963).
— Proc. Nat. Acad. Sci. US **53**, 1415 (1965).
Hershko, A., S. Amoz, and J. Mager: Biochem. Biophys. Res. Communs. **5**, 46 (1961).
Hess, E. L., and R. Horn: J. Mol. Biol. **10**, 541 (1964).
—, and S. E. Lagg: Biochemistry **2**, 726 (1963).
Huxley, H. E., and G. Zubay: J. Mol. Biol. **2**, 10 (1960).

Infante, A., and M. Nemer: J. Mol. Biol. **32**, 543 (1968).
Inouye, A., Y. Shinagawa, and S Masumura: Nature **199**, 1290 (1963).

Keller, P. J., E. Cohen, and R. D. Wade: Biochemistry **2**, 315 (1963).
Kuff, E. L., and R. F. Zeigel: J. Biophys. Biochem. Cytol. **7**, 465 (1960).
Küntzel, H., and H. Noll: Nature **215**, 1340 (1967).
Kurland, C. G.: J. Mol. Biol. **2**, 83 (1960).

Lamfrom, H., and E. R. Glowacki: J. Mol. Biol. **5**, 97 (1962).
Langridge, R., and K. C. Holmes: J. Mol. Biol. **5**, 611 (1962).
Lansink, A. G. W. J.: Yeast ribosomes and magnesium ions. Nijmegen, Netherlands: Doctor Thesis 1964.
Lerman, M. I., A. S. Spirin, L. P. Gavrilova, and V. F. Golov: J. Mol. Biol. **15**, 268 (1966).
Lucas, J. M., A. H. W. M. Schüürs, and M. V. Simpson: Federation Proc. **22**, 302 (1963).
Lyttleton, J. W.: Biochem. J. **74**, 82 (1960).
— Exp. Cell Research **26**, 312 (1962).

Madison, J. T., and S. R. Dickman: Biochemistry **2**, 321 (1963).
Martin, R. G., and B. N. Ames: Proc. Nat. Acad. Sci. US **48**, 2171 (1962).

Nathans, D., and F. Lipmann: Proc. Nat. Acad. Sci. US **47**, 497 (1961).

Odintsova, M. S., V. I. Bruskov, and E. V. Golubeva: Biokhinija **32**, 1047 (1967).
Ohtaka, Y., and K. Uchida: Biochim. et Biophys. Acta **76**, 94 (1963).

Perry, R. P., and D. E. Kelly: J. Mol. Biol. **16**, 255 (1966).
Petermann, M. L.: J. Biol. Chem. **235**, 1998 (1960).
—, and M. G. Hamilton: J. Biol. Chem. **224**, 725 (1957).
—, and A. Pavlovec: Biochim. et Biophys. Acta **114**, 264 (1966).
Pogo, A. O., B. B. Pogo, V. C. Littau, V. G. Allfrey, A. E. Mirsky, and M. G. Hamilton: Biochim. et Biophys. Acta **55**, 849 (1962).
Privalov, P. L.: Biofizika **13**, 187 (1968).
—, and G. M. Mrevlishvili: Biofizika **12**, 22 (1967).

Rodgers, A.: Biochem. J. **90**, 548 (1964).

Sager, R.: Science **157**, 709 (1967).
Shelton, E., and E. L. Kuff: J. Mol. Biol. **22**, 23 (1966).

Siekevitz, P., and G. E. Palade: J. Biophys. Biochem. Cytol. **7**, 631 (1960).
— — J. Cell. Biol. **13**, 217 (1962).
Spahr, P. F.: J. Mol. Biol. **4**, 395 (1962).
Spencer, D.: Arch. Biochim. Biophys. **111**, 381 (1965).
Spirin, A. S., and N. A. Kisselev: Communication at the VI Intern. Congress of Biochem., New York (Abstracts I, Symp. 3, p. 32). 1964.
— —, R. S. Shakulov, and A. A. Bogdanov: Biokhimiya **28**, 920 (1963).
Svetailo, E. N., I. I. Philippovich, and N. M. Sissakian: Doklady Akad. Nauk S.S.S.R. **170**, 206 (1966).
— — — J. Mol. Biol. **24**, 405 (1967).

Tashiro, Y., and P. Siekevitz: J. Mol. Biol. **11**, 149 (1965).
—, and D. A. Yphantis: J. Mol. Biol. **11**, 174 (1965).
Taylor, M. M., and R. Storck: Proc. Nat. Acad. Sci. US **52**, 958 (1964).
Tissières, A., and J. D. Watson: Nature **182**, 778 (1958).
— —, D. Schlessinger, and B. R. Hollingworth: J. Mol. Biol. **1**, 221 (1959).
Ts'o, P. O. P., and J. Vinograd: Biochim. et Biophys. Acta **49**, 113 (1961).
—, J. Bonner, and J. Vinograd: J. Biophys. Biochem. Cytol. **2**, 451 (1956).
— — — Biochim. et Biophys. Acta **30**, 570 (1958).

Zillig, W., W. Krone, and M. Albers: Hoppe-Seyler's Z. physiol. Chem. **317**, 131 (1959).

II. Ribosomal RNA

Antonov, A. S., and A. N. Belozersky: Doklady Akad. Nauk S.S.S.R. **142**, 1184 (1962).
Applebaum, S. W., R. P. Ebstein, and G. R. Wyatt: J. Mol. Biol. **21**, 29 (1966).
Aronson, A. I.: J. Mol. Biol. **5**, 453 (1962).
— Biochim. et Biophys. Acta **72**, 176 (1963).
—, and M. A. Holowczyk: Biochim. et Biophys. Acta **95**, 217 (1965).
—, and B. J. McCarthy: Biophys. J. **1**, 215 (1961).
Attardi, G., P. C. Huang, and S. Kabat: (1) Proc. Nat. Acad. Sci. US **53**, 1490 (1965).
— — — (2) Proc. Nat. Acad. Sci. US **54**, 185 (1965).

Belozersky, A. N., and A. S. Spirin: Nature **182**, 111 (1958).
— — The Nucleic Acids **3**, p. 147. New York: Acad. Press 1960.
Blake, A., and A. R. Peacocke: Nature **208**, 1319 (1965).
Boedtker, H., W. Möller, and E. Klemperer: Nature **194**, 444 (1962).
Bogdonova, E. S., L. P. Gavrilova, G. A. Dvorkin, N. A. Kisselev, and A. S. Spirin: Biokhimiya **27**, 387 (1962).
Bolton, E. T., R. J. Britten, D. B. Cowie, B. J. McCarthy, K. McQuillen, and R. B. Roberts: Carnegie Inst. Wash. Year Book **58**, 259 (1959).
Bonhoeffer, F., and H. K. Schachman: Biochem. Biophys. Res. Communs. **2**, 366 (1960).
Brown, D. D., and J. B. Gurdon: Proc. Nat. Acad. Sci. US **51**, 139 (1964).
—, and E. Littna: J. Mol. Biol. **8**, 669 (1964).
Brownlee, G. G., and F. Sanger: J. Mol. Biol. **23**, 337 (1967).
— —, and B. G. Barrel: Nature **215**, 735 (1967).

Chargaff, E.: Essays on nucleic acids. Amsterdam-London-New York: Elsevier Publ. Co. 1963.
Cheng, P.-Y.: Nature **184**, 190 (1959).
— Biochim. et Biophys. Acta **37**, 238 (1960).
Click, R. E., and D. P. Hackett: J. Mol. Biol. **17**, 279 (1966).
—, and B. L. Tint: J. Mol. Biol. **25**, 111 (1967).
Comb, D. G., and S. Katz: J. Mol. Biol. **8**, 790 (1964).
—, and N. Sarkar: J. Mol. Biol. **25**, 317 (1967).
— —, J. DeVallet, and C. J. Pinzino: J. Mol. Biol. **12**, 509 (1965).
—, and T. Zehavi-Willner: J. Mol. Biol. **23**, 441 (1967).

Cox, R. A.: Colloque Internat. sur les Acides Ribonucléiques et Polyphosphates, Strasbourg, 6-12 juillet, 1961; p. 135. Paris: CNRS 1962.
— Biochem. J. **98**, 841 (1966).

DeBellis, R. H., N. Gluck, and P. A. Marks: J. Clin. Invest. **43**, 1329 (1964).
Doi, R. H., and R. T. Igarashi: J. Bacteriol. **90**, 384 (1965).
— — J. Bacteriol. **92**, 88 (1966).
Doty, P., H. Boedtker, J. R. Fresco, R. Haselkorn, and M. Litt: Proc. Nat. Acad. Sci. US **45**, 482 (1959).
Dubin, D., and A. Günalp: Biochim. et Biophys. Acta **134**, 106 (1967).
Dubnau, D., I. Smith, P. Morell, and J. Marmur: Proc. Nat. Acad. Sci. US **54**, 491 (1965).

Ellem, K. A. O.: J. Mol. Biol. **20**, 283 (1966).
Elson, D., and E. Chargaff: Nature **173**, 1037 (1954).
— — Biochim. et Biophys. Acta **17**, 367 (1955).

Fresco, J. R., B. M. Alberts, and P. Doty: Nature **188**, 98 (1960).
Furano, A. V., D. F. Bradley, and L. G. Childers: Biochemistry **5**, 3044 (1966).

Galibert, F., C. J. Larsen, J. C. Lelong, et M. Boiron: Bull. soc. chim. biol. **48**, 21 (1966).
Gavrilova, L. P., A. S. Spirin, and A. N. Belozersky: Doklady Akad. Nauk S.S.S.R. **126**, 1121 (1959).
Gazarjan, K. G., and N. G. Schuppe: Biokhimiya **31**, 687 (1966).
— — Doklady Akad. Nauk S.S.S.R. **176**, 714 (1967).
Gierer, A.: Z. Naturforsch. **13 B**, 788 (1958).
Green, M., and B. Hall: Biophys. J. **1**, 517 (1961).
Gumilevskaja, N. A., and N. M. Sissakian: Doklady Akad. Nauk S.S.S.R. **149**, 198 (1963).

Hall, B. D., and P. Doty: Microsomal particles and protein synthesis, p. 27. Oxford: Pergamon Press 1958.
— — J. Mol. Biol. **1**, 111 (1959).
Hamilton, M.: Biochim. et Biophys. Acta **134**, 473 (1967).
Haselkorn, R.: J. Mol. Biol. **4**, 357 (1962).
Hayashi, Y., S. Osawa, and K. Miura: Biochim. et Biophys. Acta **129**, 519 (1966).
Helmkamp, G. K., and P. O. P. Ts'o: J. Am. Chem. Soc. **83**, 138 (1961).
Henshaw, E. C.: J. Mol. Biol. **9**, 610 (1964).
Hirsch, C. A.: Biochim. et Biophys. Acta **123**, 246 (1966).
Huxley, H. E., and G. Zubay: J. Mol. Biol. **2**, 10 (1960).

Kisselev, N. A., L. P. Gavrilova, and A. S. Spirin: J. Mol. Biol. **3**, 778 (1961).
Klug, A., K. C. Holmes, and J. T. Finch: J. Mol. Biol. **3**, 87 (1961).
Kronman, M. J., S. N. Timasheff, J. S. Colter, and R. A. Brown: Biochim. et Biophys. Acta **40**, 410 (1960).
Küntzel, H., and H. Noll: Nature **215**, 1340 (1967).
Kurland, C. G.: J. Mol. Biol. **2**, 83 (1960).

Lane, B. G., J. Diemer, and C. A. Blashko: Can. J. Biochem. and Physiol. **41**, 1927 (1963).
Lerner, A. M., E. Bell, and J. E. Darnell: Science **141**, 1187 (1963).
Littauer, U. Z.: Protein biosynthesis. New York: Acad. Press 1961.

McPhie, P., and W. B. Gratzer: Biochemistry **5**, 1310 (1966).
Mednikov, B. M.: Doklady Akad. Nauk S.S.S.R. **161**, 721 (1965).
—, A. S. Antonov, and A. N. Belozersky: Doklady Akad. Nauk S.S.S.R. **165**, 227 (1965).
Midgley, J. E. M.: Biochim. et Biophys. Acta **61**, 513 (1962).
— (1) Biochim. et Biophys. Acta **108**, 340 (1965).
— (2) Biochim. et Biophys. Acta **108**, 348 (1965).
Möller, W., and H. Boedtker: Colloque internat. sur les acides ribonucléiques et polyphosphates. Strasbourg, 6-12 juillet 1961, p. 99. Paris: CNRS 1962.
Monier, R.: Lecture at the FEBS Summer School on nucleic acids, Marseilles, France 1967.
Montagnier, L., and A. D. Bellamy: Biochim. et Biophys. Acta **80**, 157 (1964).

Oishi, M., and N. Sueoka: Proc. Nat. Acad. Sci. US **54**, 483 (1965).

OPARA-KUBINSKA, Z., H. KUBINSKI, and W. SZYBALSKI: Proc. Nat. Acad. Sci. US **52**, 923 (1964).
OSAWA, S.: Biochim. et Biophys. Acta **43**, 110 (1960).
PERMOGOROV, V. I., and I. A. SLADKOVA: Molekularnaya biologia (U.S.S.R.) **2**, 276 (1968).
PETERMANN, M. L., and A. PAVLOVEC: J. Biol. Chem. **238**, 3717 (1963).
— — Biochim. et Biophys. Acta **114**, 264 (1966).
POLLARD, C. J.: Biochem. Biophys. Res. Communs. **17**. 171 (1964),

PRESTAYKO, A. W., and W. D. FISHER: J. Cell Biol. **31**, 88 A (1966).

RITOSSA, F. M., and S. SPIEGELMAN: Proc. Nat. Acad. Sci. US **53**, 737 (1965).
—, K. C. ATWOOD, and S. SPIEGELMAN: Genetics **54**, 663 (1966).
ROSSET, R., and R. MONIER: Biochim. et Biophys. Acta **68**, 653 (1963).
— —, et J. JULIEN: Bull. soc. chim. biol. **46**, 87 (1964).

SANGER, F., G. G. BROWNLEE, and B. G. BARRELL: J. Mol. Biol. **13**, 373 (1965).
SANTER, M.: Science **141**, 1049 (1963).
SARKAR, P. K., J. T. YANG, and P. DOTY: Biopolymers **5**, 1 (1967).
SCHLEICH, T., and J. GOLDSTEIN: J. Mol. Biol. **15**, 136 (1966).
SCHLESSINGER, D.: J. Mol. Biol. **2**, 92 (1960).
SHAKULOV, R. S., M. A. AJTKHOZHIN, and A. S. SPIRIN: Biokhimiya **27**, 744 (1962).
SPAHR, P. F., and A. TISSIÈRES: J. Mol. Biol. **1**, 237 (1959).
SPENCER, M., and F. POOLE: J. Mol. Biol. **11**, 314 (1965).
—, W. FULLER, M. H. F. WILKINS, and G. L. BROWN: Nature **194**, 1014 (1962).
SPIRIN, A. S.: J. Mol. Biol. **2**, 436 (1960).
— Biokhimiya **26**, 511 (1961).
— Colloque intern. sur les acides ribonucléiques et les polyphosphates. Strasbourg, 6-12 juillet 1961, p. 73. Paris: CNRS 1962.
— Some problems concerning the macromolecular structure of ribonucleic acids (in Russian). Akad. Nauk S.S.S.R., Moscow. Translation: Macromolecular structure of ribonucleic acids, 1963. New York: Reinhold 1964.
—, and L. S. MILMAN: Doklady Akad. Nauk S.S.S.R. **134**, 717 (1960).
—, A. N. BELOZERSKY, N. V. SHUGAYEVA, and B. F. VANYUSHIN: Biokhimiya **22**, 744 (1957).
—, L. P. GAVRILOVA, S. E. BRESLER, and M. I. MOSEVITSKIY: Biokhimiya **24**, 938 (1959).
—, N. A. KISSELEV, R. S. SHAKULOV, and A. A. BOGDANOV: Biokhimiya **28**, 920 (1963).
STANLEY, W. M., and R. M. BOCK: Biochemistry **4**, 1302 (1965).
STUTZ, E., and H. NOLL: Proc. Nat. Acad. Sci. US **57**, 774 (1967).

TAKANAMI, M.: Biochim. et Biophys. Acta **39**, 152 (1960).
TASHIRO, Y., H. SHIMIDZU, A. INOUYE, and K. KAKIUCHI: Biochim. et Biophys. Acta **43**, 544 (1960).
TAYLOR, M. M., J. E. GLASGOW, and R. STORCK: Proc. Nat. Acad. Sci. US **57**, 164 (1967).
TIMASHEFF, S. N., R. A. BROWN, J. S. COLTER, and M. DAVIES: Biochim. et Biophys. Acta **27**, 662 (1958).
TS'O, P. O. P., and G. K. HELMKAMP: Tetrahedron **13**, 198 (1961).
—, and R. SQUIRES: Fed. Proc. **18**, 341 (1959).

URYSON, S. O., and A. N. BELOZERSKY: Doklady Akad. Nauk S.S.S.R. **125**, 1144 (1959).
— — Doklady Akad. Nauk S.S.S.R. **132**, 708 (1960).

VANYUSHIN, B. F., and A. N. BELOZERSKY: Doklady Akad. Nauk S.S.S.R. **127**, 455 (1959).
— —, and S. L. BOGDANOVA: Doklady Akad. Nauk S.S.S.R. **134**, 1222 (1960).

WOESE, C. R.: Nature **189**, 920 (1961).

YANKOFSKY, S. A., and S. SPIEGELMAN: Proc. Nat. Acad. Sci. US **48**, 1466 (1962).
— — Proc. Nat. Acad. Sci. US **49**, 538 (1963).

ZEHAVI-WILLNER, T., and D. G. COMB: J. Mol. Biol. **16**, 250 (1966).
ZIMMERMAN, E. F., M. HEETER, and J. E. DARNELL: Virology **19**, 400 (1963).
ZUBAY, G., and M. H. F. WILKINS: J. Mol. Biol. **2**, 105 (1960).

III. Ribosomal Proteins

ATSMON, A., P. SPITNIK-ELSON, and D. ELSON: J. Mol. Biol. **25**, 161 (1967).

BAYLEY, S. T.: J. Mol. Biol. **15**, 420 (1966).

CAMMACK, K. A., and H. E. WADE: Biochem. J. **96**, 671 (1965).
CRAMPTON, C. F., and M. L. PETERMANN: J. Biol. Chem. **234**, 2642 (1959).
CURRY, J. B., and R. T. HERSH: Biochem. Biophys. Res. Communs. **6**, 415 (1962).

ELSON, D., and M. TAL: Biochim. et Biophys. Acta **36**, 281 (1959).

GARBER, M. B., and N. A. BARULINA: Molekularnaya Biologia (U.S.S.R.) **2**, 448 (1968).
GESTELAND, R. F.: (1) J. Mol. Biol. **16**, 67 (1966).
— (2) J. Mol. Biol. **18**, 356 (1966).

HAMILTON, M., and M. RUTH: Biochemistry **6**, 2585 (1967).

KALTSCHMIDT, E., M. DZIONARA, D. DONNER, and H. G. WITTMANN: Molec. Gen. Genetics *100*, 364 (1967).

LEBOY, P. S., E. C. COX, and J. G. FLAKS: Proc. Nat. Acad. Sci. US **52**, 1367 (1964).
LERMAN, M. I., A. S. SPIRIN, L. P. GAVRILOVA, and V. F. GOLOV: J. Mol. Biol. **15**, 268 (1966).
LOW, R. B., and I. G. WOOL: Science **155**, 330 (1967).

MARCOT-QUEIROZ, J., et R. MONIER: Bull. soc. chim. biol. **48**, 446 (1966).
MATHIAS, A. P., and R. WILLIAMSON: J. Mol. Biol. **9**, 498 (1964).
MCPHIE, P., and W. B. GRATZER: Biochemistry **5**, 1310 (1966).
MESELSON, M., M. NOMURA, S. BRENNER, C. DAVERN, and D. SCHLESSINGER: J. Mol. Biol. **9**, 696 (1964).
MÖLLER, W., and A. CHRAMBACH: J. Mol. Biol. **23**, 377 (1967).
—, and J. WIDDOWSON: J. Mol. Biol. **24**, 367 (1967).
MOORE, P. B., R. R. TRAUT, H. NOLLER, P. PEARSON, and H. DELIUS: J. Mol. Biol. **31**, 441 (1968).

NEU, H. C., and L. A. HEPPEL: J. Biol. Chem. **239**, 2927 (1964).
NOMURA, M., and P. TRAUB: Organizational Biosynthesis, New York: Acad. Press 1966.

PETERMANN, M. L.: The physical and chemical properties of ribosomes. Amsterdam-London-New York: Elsevier Publ. Co. 1964.

SALAS, M., M. A. SMITH, W. M. STANLEY, A. J. WAHBA, and S. OCHOA: J. Biol. Chem. **240**, 3988 (1965).
SARKAR, P. K., J. T. YANG, and P. DOTY: Biopolymers **5**, 1 (1967).
SETTERFIELD, G., J. M. NEELIN, E. M. NEELIN, and S. T. BAYLEY: J. Mol. Biol. **2**, 416 (1960).
SPAHR, P. F.: J. Mol. Biol. **4**, 395 (1962).
SPIRIN, A. S., N. V. BELITSINA, and M. I. LERMAN: J. Mol. Biol. **14**, 611 (1965).
—, N. A. KISSELEV, R. S. SHAKULOV, and A. A. BOGDANOV: Biokhimiya **28**, 920 (1963).
SPITNIK-ELSON, P.: Biochim. et Biophys. Acta **80**, 594 (1964).

TRAUB, P., M. NOMURA, and L. TU: J. Mol. Biol. **19**, 215 (1966).
TRAUT, R.: J. Mol. Biol. **21**, 571 (1966).
—, P. MOORE, H. DELIUS, H. NOLLER, and A. TISSIÈRES: Proc. Nat. Acad. Sci. US **57**, 1294 (1967).
TS'O, P. O. P., J. BONNER, and H. DINTZIS: Arch. Biochem. Biophys. **76**, 225 (1958).

WALLER, J. P.: J. Mol. Biol. **7**, 483 (1963).
— J. Mol. Biol. **10**, 319 (1964).
—, and J. I. HARRIS: Proc. Nat. Acad. Sci. US **47**, 18 (1961).

IV. Structural Transformations of Ribosomes

BAYLEY, S. T.: J. Mol. Biol. **15**, 420 (1966).
—, and D. J. KUSHNER: J. Mol. Biol. **9**, 654 (1964).
BRITTEN, R. J., B. J. MCCARTHY, and R. B. ROBERTS: Biophys. J. **2**, 83 (1962).

CAMMACK, K. A., and H. E. WADE: Biochem. J. **96**, 671 (1965).
CHAO, F. C.: Arch. Biochem. Biophys. **70**, 426 (1957).
COHEN, S. S., and J. LICHTENSTEIN: J. Biol. Chem. **235**, 2112 (1960).

DAHLBERG, J. E., and R. HASELKORN: J. Mol. Biol. **24**, 83 (1967).

ELSON, D.: Biochim. et Biophys. Acta **53**, 232 (1961).
— Biochim. et Biophys. Acta **80**, 379 (1964).

GAVRILOVA, L. P., D. A. IVANOV, and A. S. SPIRIN: (1) J. Mol. Biol. **16**, 473 (1966).
—, M. I. LERMAN, and A. S. SPIRIN: (2) Izvest. Akad. Nauk S.S.S.R. Ser. Biol. N **6**, 826 (1966).
GESTELAND, R. F.: J. Mol. Biol. **18**, 356 (1966).
GORDON, J., and F. LIPMANN: J. Mol. Biol. **23**, 23 (1967).
GOULD, H. J., H. R. V. ARNSTEIN, and R. A. COX: J. Mol. Biol. **15**, 600 (1966).
GREEN, M., and B. HALL: Biophys. J. **1**, 517 (1961).

HAMILTON, M. G., and M. L. PETERMANN: J. Biol. Chem. **234**, 1441 (1959).
HOSOKAWA, K., R. K. FUJIMURA, and M. NOMURA: Proc. Nat. Acad. Sci. US **55**, 198 (1966).
HUXLEY, H. E., and G. ZUBAY: J. Mol. Biol. **2**, 10 (1960).

KUFF, E. L., and R. F. ZEIGEL: J. Biophys. Biochem. Cytol. **7**, 465 (1960).
KURLAND, C. G., M. NOMURA, and J. D. WATSON: J. Mol. Biol. **4**, 388 (1962).

LAMFROM, H., and E. R. GLOWACKI: J. Mol. Biol. **5**, 97 (1962).
LEDERBERG, S., and J. M. MITCHISON: Biochim. et Biophys. Acta **55**, 104 (1962).
LERMAN, M. I.: Molekularnaya Biologia (U.S.S.R.) **2**, 209 (1968).
—, A. S. SPIRIN, L. P. GAVRILOVA, and V. F. GOLOV: J. Mol. Biol. **15**, 268 (1966).
—, R. A. ZIMMERMANN, L. P. GAVRILOVA, and A. S. SPIRIN: Molekularnaya Biologia (U.S.S.R.) **1**, 12 (1967).

MANGIAROTTI, G., and D. SCHLESSINGER: J. Mol. Biol. **20**, 123 (1966).
MARCOT-QUEIROZ, J., et R. MONIER: (1) Bull. soc. chim. biol. **47**, 1627 (1965).
— — (2) J. Mol. Biol. **14**, 490 (1965).
— — Bull. soc. chim. biol. **48**, 446 (1966).
McPHIE, P., and W. B. GRATZER: Biochemistry **5**, 1310 (1966).
MESELSON, M., M. NOMURA, S. BRENNER, C. DAVERN, and D. SCHLESSINGER: J. Mol. Biol. **9**, 696 (1964).
MOORE, P. B.: J. Mol. Biol. **22**, 145 (1966).
MORGAN, R. S.: J. Mol. Biol. **4**, 115 (1962).
—, C. GREENSPAN, and B. CUNNINGHAM: Biochim. et Biophys. Acta **68**, 642 (1963).

NAKADA, D., and A. KAJI: Biochemistry **57**, 128 (1967).
—, and J. UNOWSKY: Proc. Nat. Acad. Sci. US **56**, 659 (1966).
NOMURA, M., and P. TRAUB: Organizational Biosynthesis, New York: Acad. Press 1966.
—, and J. D. WATSON: J. Mol. Biol. **1**, 204 (1959).

OSAWA, S., E. OTAKA, A. MUTO, K. YOSHIDA, and T. ITOH: Communication at the 7th Intern. Congress of Biochem., Tokyo, Japan. (Abstracts I, Symp. II-3,6, p. 119). 1967.

PESTKA, S.: J. Biol. Chem. **241**, 367 (1966).
PETERMANN, M. L.: J. Biol. Chem. **235**, 1998 (1960).
—, and A. PAVLOVEC: Biochim. et Biophys. Acta **114**, 264 (1966).

RASKAS, H. J., and T. STAEHELIN: J. Mol. Biol. **23**, 89 (1967).
RODGERS, A.: Biochem. J. **90**, 548 (1964).

SARKAR, P. K., J. T. YANG, and P. DOTY: Biopolymers **5**, 1 (1967).
SCHLESSINGER, D., G. MANGIAROTTI, and D. APIRION: Proc. Nat. Acad. Sci. US **58**, 1782 (1967).
SHAKULOV, R. S., A. A. BOGDANOV, and A. S. SPIRIN: Doklady Akad. Nauk S.S.S.R. **153**, 224 (1963).
SPIRIN, A. S.: Cold Spring Harbor Symposia Quant. Biol. **28**, 267 (1963).

SPIRIN, A. S., and N. V. BELITSINA: J. Mol. Biol. **15**, 282 (1966).
— —, Ö. GAAL, and T. M. POZDNYAKOVA: Molekularnaya Biologia (U.S.S.R.) **2**, 95 (1968).
—, H. V. BELITSINA, and M. I. LERMAN: J. Mol. Biol. **14**, 611 (1965).
—, N. A. KISSELEV, R. S. SHAKULOV, and A. A. BOGDANOV: Biokhimiya **28**, 920 (1963).
—, M. I. LERMAN, L. P. GAVRILOVA, and N. V. BELITSINA: Biokhimiya **31**, 424 (1966).
STAEHELIN, T., and M. MESELSON: J. Mol. Biol. **16**, 245 (1966).

TAKEDA, M., and F. LIPMANN: Proc. Nat. Acad. Sci. US **56**, 1875 (1966).
TAMAOKI, T., and F. MIYAZAWA: J. Mol. Biol. **23**, 35 (1967).
TASHIRO, Y., and P. SIEKEVITZ: J. Mol. Biol. **11**, 149 (1965).
TISSIÈRES, A., and J. D. WATSON: Nature **182**, 778 (1958).
— —, D. SCHLESSINGER, and B. R. HOLLINGWORTH: J. Mol. Biol. **1**, 221 (1959).
TS'O, P. O. P.: Microsomal particles and protein synthesis, p. 156. Oxford: Pergamon Press 1958.
—, and J. VINOGRAD: Biochim. et Biophys. Acta **49**, 113 (1961).
—, J. BONNER, and J. VINOGRAD: Biochim. et Biophys. Acta **30**, 570 (1958).

WATSON, J. D.: Bull. soc. chim. biol. **46**, 1399 (1964).
WELLER, D. L., and J. HOROWITZ: Biochim. et Biophys. Acta **87**, 361 (1964).

ZAK, R., K. NAIR, and M. RABINOWITZ: Nature **210**, 169 (1966).

Part Two

I. Components of the Protein-Synthesizing System

ALLENDE, J. E., R. MONRO, and F. LIPMANN: Proc. Nat. Acad. Sci. US **51**, 1211 (1964).
—, N. W. SEEDS, T. W. CONWAY, and H. WEISSBACH: Proc. Nat. Acad. Sci. US **58**, 1566 (1967).
ARLINGHAUS, R., J. SHAEFFER, and R. SCHWEET: Proc. Nat. Acad. Sci. US **51**, 1291 (1964).

BAUTZ, E. K. F.: Cold Spring Harbor Symposia Quant. Biol. **28**, 205 (1963).
BISHOP, J. O., and R. S. SCHWEET: Biochim. et Biophys. Acta **54**, 617 (1961).
BRAWERMAN, G., and J. EISENSTADT: Biochemistry **5**, 2784 (1966).

CAPECCHI, M. R.: Proc. Nat. Acad. Sci. US **58**, 1144 (1967).
CHAN, M., and D. J. MCCORQUODALE: J. Biol. Chem. **240**, 3116 (1965).
CONWAY, T. W.: Proc. Nat. Acad. Sci. US **51**, 1216 (1964).
—, and F. LIPMANN: Proc. Nat. Acad. Sci. US **52**, 1462 (1964).

VON EHRENSTEIN, G., and F. LIPMANN: Proc. Nat. Acad. Sci. US **47**, 941 (1961).

FESSENDEN, J. M., and K. MOLDAVE: Biochem. Biophys. Res. Communs. **6**, 232 (1961).
— — Biochemistry **1**, 485 (1962).
— — J. Biol. Chem. **238**, 1479 (1963).

GANOZA, M. C.: Cold Spring Harbor Symposia Quant. Biol. **31**, 273 (1966).
GILBERT, W.: J. Mol. Biol. **6**, 374 (1963).
GORDON, J.: Proc. Nat. Acad. Sci. US **58**, 1574 (1967).
—, and F. LIPMANN: J. Mol. Biol. **23**, 23 (1967).
GROS, F.: Lecture at the FEBS Summer School on nucleic acids, Marseilles. 1967.

HASELKORN, R., V. A. FRIED, and J. E. DAHLBERG: Proc. Nat. Acad. Sci. US **49**, 511 (1963).
HOAGLAND, M.: The Nucleic Acids **3**, 349. New York: Acad. Press 1960.
HOLLEY, R. W., J. APGAR, G. A. EVERETT, J. T. MADISON, M. MARQUISEE, S. H. MERRILL, J. R. PENSWICK, and A. ZAMIR: Science **147**, 1462 (1965).

IBUKI, F., E. GASIOR, and K. MOLDAVE: J. Biol. Chem. **241**, 2188 (1966).

KAJI, H., and A. KAJI: Fed. Proc. **24**, 408 (1965).
KELLER, E. B., and P. C. ZAMECNIK: J. Biol. Chem. **221**, 45 (1956).
KUĆAN, Ž.: Radiation Research **27**, 229 (1966).

LERMAN, M. I., A. S. SPIRIN, L. P. GAVRILOVA, and V. F. GOLOV: J. Mol. Biol. **15**, 268 (1966).
LUBIN, M.: Biochim. et Biophys. Acta **72**, 345 (1963).
— Fed. Proc. **23**, 994 (1964).
—, and H. L. ENNIS: Fed. Proc. **22**, 302 (1963).
— — Biochim. et Biophys. Acta **80**, 614 (1964).
LUCAS-LENARD, J., and F. LIPMANN: Proc. Nat. Acad. Sci. US **55**, 1562 (1966).

MCCARTHY, B. J., and J. J. HOLLAND: Proc. Nat. Acad. Sci. US **54**, 880 (1965).
MOORE, P. B.: J. Mol. Biol. **18**, 8 (1966).
MORGAN, R., R. D. WELLS, and H. G. KHORANA: J. Mol. Biol. **26**, 477 (1967).

NAKAMOTO, T., T. W. CONWAY, J. E. ALLENDE, G. J. SPYRIDES, and F. LIPMANN: Cold Spring Harbor Symposia Quant. Biol. **28**, 227 (1963).
NATHANS, D., and F. LIPMANN: Biochim. et Biophys. Acta **43**, 126 (1960).
— — Proc. Nat. Acad. Sci. US **47**, 497 (1961).
—, G. VON EHRENSTEIN, R. MONRO, and F. LIPMANN: Fed. Proc. **21**, 127 (1962).
NIRENBERG, M. W., and J. H. MATTHAEI: Proc. Nat. Acad. Sci. US **47**, 1588 (1961).
—, O. W. JONES, P. LEDER, B. F. C. CLARK, W. S. SLY, and S. PESTKA: Cold Spring Harbor Symposia Quant. Biol. **28**, 549 (1963).
NISHIZUKA, Y., and F. LIPMANN: (1) Proc. Nat. Acad. Sci. US **55**, 212 (1966).
— — (2) Arch. Biochem. Biophys. **116**, 344 (1966).
NOMURA, M., and P. TRAUB: Organizational Biosynthesis, New York: Acad. Press 1966.

PESTKA, S., and M. NIRENBERG: Cold Spring Harbor Symposia Quant. Biol. **31**, 641 (1966).

RASKAS, H., and T. STAEHELIN: J. Mol. Biol. **23**, 89 (1967).
REVEL, M., and F. GROS: Biochem. Biophys. Res. Communs. **25**, 124 (1966).
— — Biochem. Biophys. Res. Communs. **27**, 12 (1967).
RYCHLIK, I.: Biochim. et Biophys. Acta **114**, 425 (1966).
—, S. CHLÁDEK, and J. ŽEMLIČKA: Biochim. et Biophys. Acta **138**, 640 (1967).

SALAS, M., M. B. HILLE, J. A. LAST, A. J. WAHBA, and S. OCHOA: (1) Proc. Nat. Acad. Sci. US **57**, 387 (1967).
—, M. J. MILLER, A. J. WAHBA, and S. OCHOA: (2) Proc. Nat. Acad. Sci. US **57**, 1865 (1967).
SHAKULOV, R. S., M. A. AJTKHOZHIN, and A. S. SPIRIN: Biokhimiya **27**, 744 (1962).
SINGER, M. F., O. W. JONES, and M. W. NIRENBERG: Proc. Nat. Acad. Sci. US **49**, 392 (1963).
SMITH, J. D., R. R. TRAUT, G. M. BLACKBURN, and R. E. MONRO: J. Mol. Biol. **13**, 617 (1965).
SPEYER, J. F., P. LENGYEL, C. BASILIO, A. J. WAHBA, R. S. GARDNER, and S. OCHOA: Cold Spring Harbor Symposia Quant. Biol. **28**, 559 (1963).
SPIRIN, A. S.: Some problems concerning the macromolecular structure of ribonucleic acids (in Russian). Izd. Akad. Nauk S.S.S.R., Moscow. Translation: Macromolecular structure of ribonucleic acids, 1963. New York: Reinhold 1964.
SPYRIDES, G. J.: Proc. Nat. Acad. Sci. US **51**. 1220 (1964).
STANLEY, W. M., M. SALAS, A. J. WAHBA, and S. OCHOA: Proc. Nat. Acad. Sci. US **56**, 290 (1966).
SUTTER, R., and K. MOLDAVE: J. Biol. Chem. **241**, 1698 (1966).
SZER, W., and S. OCHOA: J. Mol. Biol. **8**, 823 (1964).

TAKANAMI, M., and T. OKAMOTO: (1) J. Mol. Biol. **7**, 323 (1963).
— — (2) Biochem. Biophys. Res. Communs. **13**, 297 (1963).
TRAUT, R. R., and R. E. MONRO: J. Mol. Biol. **10**, 63 (1964).

ZAK, R., K. NAIR, and M. RABINOWITZ: Nature **210**, 169 (1966).

II. Association of the Ribosome with Components of the Protein-Synthesizing System

CANNON, M.: Biochem. J. **104**, 934 (1967).
—, R. KRUG, and W. GILBERT: J. Mol. Biol. **7**, 360 (1963).

CHAPEVILLE, F., F. LIPMANN, G. VON EHRENSTEIN, B. WEISBLUM, W. J. RAY, and S. BENZER: Proc. Nat. Acad. Sci. US **48**, 1086 (1962).
CONWAY, T. W.: Proc. Nat. Acad. Sci. US **51**, 1216 (1964).
CRICK, F. H. C.: Symposium Soc. Exptl. Biol. **12**, 138 (1958).

DAHLBERG, J. E., and R. HASELKORN: J. Mol. Biol. **24**, 83 (1967).

HASELKORN, R., and V. A. FRIED: Proc. Nat. Acad. Sci. US **51**, 1001 (1964).
— —, and J. E. DAHLBERG: Proc. Nat. Acad. Sci. US **49**, 511 (1963).
HATFIELD, D.: Fed. Proc. **24**, 409 (1965).
— Cold Spring Harbor Symposia Quant. Biol. **31**, 619 (1966).

KAJI, A., and H. KAJI: Biochem. Biophys. Res. Communs. **13**, 186 (1963).
KAJI, H., and A. KAJI: (1) Proc. Nat. Acad. Sci. US **52**, 1541 (1964).
KAJI, A., and H. KAJI: (2) Biochim. et Biophys. Acta **87**, 519 (1964).
KAJI, H., I. SUZUKA, and A. KAJI: (1) J. Mol. Biol. **18**, 219 (1966).
— — — (2) J. Biol. Chem. **241**, 1251 (1966).
KEPES, A., Biochim. et Biophys. Acta **138**, 107 (1967).
KURLAND, C. G.: J. Mol. Biol. **18**, 90 (1966).

LEDER, P., and M. W. NIRENBERG: (1) Proc. Nat. Acad. Sci. US **52**, 420 (1964).
— — (2) Proc. Nat. Acad. Sci. US **52**, 1521 (1964).
LEVIN, J. G.: Fed. Proc. **25**, 778 (1966).
LIPSETT, M. N.: J. Biol. Chem. **239**, 1256 (1964).
—, L. A. HEPPEL, and D. F. BRADLEY: J. Biol. Chem. **236**, 857 (1961).

MARTIN, R. G., D. F. SILBERT, D. W. E. SMITH, and H. WHITFIELD: J. Mol. Biol. **21**, 357 (1966).
MATTHAEI, H., F. AMELUNXEN, K. ECKERT, and G. HELLER: Ber. Bunsengesellschaft Phys. Chemie **68**, 735 (1964).
MCLAUGHLIN, C., J. DONDON, M. GRUNBERG-MANAGO, A. MICHELSON, and G. SAUNDERS: Cold Spring Harbor Symposia Quant. Biol. **31**, 601 (1966).
MICHELSON, A. M.: Bull. soc. chim. biol. **47**, 1553 (1965).
MILLAR, D. B. S., III, R. CUKIER, and M. W. NIRENBERG: Biochemistry **4**, 976 (1965).
MOORE, P. B.: (1) J. Mol. Biol. **18**, 8 (1966).
(2) J. Mol. Biol. **22**, 145 (1966).
—, and K. ASANO: J. Mol. Biol. **18**, 21 (1966).

NAKAMOTO, T., T. W. CONWAY, J. E. ALLENDE, G. J. SPYRIDES, and F. LIPMANN: Cold Spring Harbor Symposia Quant. Biol. **28**, 227 (1963).
NIRENBERG, M. W., and P. LEDER: Science **145**, 1399 (1964).
NISHIZUKA, Y., and F. LIPMANN: Arch. Biochem. Biophys. **116**, 344 (1966).
NOMURA, M., and P. TRAUB: Organizational Biosynthesis, New York: Acad. Press 1966.

OKAMOTO, T., and M. TAKANAMI: Biochim. et Biophys. Acta **68**, 325 (1963).

PESTKA, S.: J. Biol. Chem. **241**, 367 (1966).
—, and M. W. NIRENBERG: (1) J. Mol. Biol. **21**, 145 (1966).
— — (2) Cold Spring Harbor Symposia Quant. Biol. **31**, 641 (1966).
RASKAS, H. J., and I. STAEHELIN: J. Mol. Biol. **23**, 89 (1967).

ROTTMAN, F., and M. W. NIRENBERG: J. Mol. Biol. **21**, 555 (1966).
RYCHLÍK, I.: Collection Czechoslov. Chem. Commun. **31**, 2583 (1966).

SEEDS, N. W., and T. W. CONWAY: Biochem. Biophys. Res. Communs. **23**, 111 (1966).
SPYRIDES, G. J.: Proc. Nat. Acad. Sci. US **51**, 1220 (1964).
—, and F. LIPMANN: Proc. Nat. Acad. Sci. US **48**, 1977 (1962).
SUZUKA, I., H. KAJI, and A. KAJI: Biochem. Biophys. Res. Communs. **21**, 187 (1965).
— — — Proc. Nat. Acad. Sci. US **55**, 1483 (1966).
SZER, W., and L. NOWAK: J. Mol. Biol. **24**, 333 (1967).

Takanami, M. and T. Okamoto: (1) J. Mol. Biol. **7**, 323 (1963).
— — (2) Biochem. Biophys. Res. Communs. **13**, 297 (1963).
—, and G. Zubay: Proc. Nat. Acad. Sci. US **51**, 834 (1964).
—, Y. Yan, and T. H. Jukes: J. Mol. Biol. **12**, 761 (1965).

Voorma, H. O., P. W. Gout, J. Van Duin, B. W. Hoogendam, and L. Bosch: Biochim. et Biophys. Acta **87**, 693 (1964).

Warner, J., M. J. Madden, and J. E. Darnell: Virology **19**, 393 (1963).

Yanofsky, C., and J. Ito: J. Mol. Biol. **21**, 313 (1966).

Zak, R., K. Nair, and M. Rabinowitz: Nature **210**, 169 (1966).

III. Stages of Translation

Adams, J. M., and M. R. Capecchi: Proc. Nat. Acad. Sci. US **55**, 147 (1966).
Allende, J. E., and H. Weissbach: Biochem. Biophys. Res. Communs. **28**, 82 (1967).
—, R. Monro, and F. Lipmann: Proc. Nat. Acad. Sci. US **51**, 1211 (1964).
Ames, B. N., and P. E. Hartman: Cold Spring Harbor Symposia Quant. Biol. **28**, 349 (1963).
Anderson, J. S., M. S. Bretscher, B. F. C. Clark, and K. A. Marcker: Nature **215**, 490 (1967).

Bauerle, R. H., and P. Margolin: Proc. Nat. Acad. Sci. US **56**, 111 (1966).
Bishop, J. O., J. Leahy, and R. S. Schweet: Proc. Nat. Acad. Sci. US **46**, 1030 (1960).
Brenner, S., and J. R. Beckwith: J. Mol. Biol. **13**, 629 (1965).
—, A. O. W. Stretton, and S. Kaplan: Nature **206**, 994 (1965).
—, L. Barnett, E. R. Katz, and F. H. C. Crick: Nature **213**, 449 (1967).
Bresler, S., R. Grajevskaja, S. Kirilov, E. Saminski, and F. Shutov: Biochim. et Biophys. Acta **123**, 534 (1966).
Bretscher, M. S.: J. Mol. Biol. **7**, 446 (1963).
— J. Mol. Biol. **12**, 913 (1965).
—, and K. A. Marcker: Nature **211**, 380 (1966).
—, H. M. Goodman, J. R. Menninger, and J. D. Smith: J. Mol. Biol. **14**, 634 (1965).
Brimacombe, R., J. Trupin, M. Nirenberg, P. Leder, M. Bernfield, and T. Jaouni: Proc. Nat. Acad. Sci. US **54**, 954 (1965).

Cannon, M.: Biochem. J. **104**, 934 (1967).
Capecchi, M. R.: (1) Proc. Nat. Acad. Sci. US **55**, 1517 (1966).
— (2) J. Mol. Biol. **21**, 173 (1966).
— Proc. Nat. Acad. Sci. US **58**, 1144 (1967).
Clark, J. M., A. Y. Chang, S. Spiegelman, and M. E. Reichmann: Proc. Nat. Acad. Sci. US **54**, 1193 (1965).
Clark, B. F. C., and K. A. Marcker: (1) J. Mol. Biol. **17**, 394 (1966).
— — (2) Nature **211**, 378 (1966).
Conway, T. W., and F. Lipmann: Proc. Nat. Acad. Sci. US **52**, 1462 (1964).
Creighton, T. E., and C. Yanofsky: J. Biol. Chem. **241**, 980 (1966).

Dintzis, H.: Science **47**, 247 (1961).

Eisenstadt, J. M., and G. Brawerman: Proc. Nat. Acad. Sci. US **58**, 1560 (1967).

Ganoza, M. C., and T. Nakamoto: Proc. Nat. Acad. Sci. US **55**, 162 (1966).
Ghosh, H. P., D. Söll, and H. G. Khorana: J. Mol. Biol. **25**, 275 (1967).
Gierer, A.: J. Mol. Biol. **6**, 148 (1963).
Gilbert, W.: (1) Cold Spring Harbor Symposia Quant. Biol. **28**, 287 (1963).
— (2) J. Mol. Biol. **6**, 389 (1963).
Gros, F.: Personal Communication: Lecture at the FEBS Summer School on nucleic acids, Marseilles, France 1967.

Hershey, J. W., and R. E. Thach: Proc. Nat. Acad. Sci. US **57**, 759 (1967).
Hille, M. B., M. J. Miller, K. Iwasaki, and A. J. Wahba: Proc. Nat. Acad. Sci. US **58**, 1652 (1967).

Ito, J., and I. P. Crawford: Genetics **52**, 1303 (1965).

Kellogg, D. A., B. P. Doctor, J. E. Loebel, and M. W. Nirenberg: Proc. Nat. Acad. Sci. US **55**, 912 (1966).
Kepes, A.: Biochim. et Biophys. Acta **138**, 107 (1967).
Khorana, H. G., H. Buchi, H. Ghosh, N. Gupta, T. M. Jacob, H. Kossel, R. Morgan, S. A. Narang, E. Ohtsuka, and R. D. Wells: Cold Spring Harbor Symposia Quant. Biol. **31**, 39 (1966).
Kolakofsky, D., and T. Nakamoto: Proc. Nat. Acad. Sci. US **56**, 1786 (1966).

Lamfrom, H., C. S. McLaughlin, and A. Sarabhai: J. Mol. Biol. **22**, 355 (1966).
Last, J. A., W. M. Stanley, M. Salas, M. B. Hille, A. J. Wahba, and S. Ochoa: Proc. Nat. Acad. Sci. US **57**, 1062 (1967).
Leder, P., and M. Nau: Proc. Nat. Acad. Sci. US **58**, 774 (1967).
Lipmann, F.: Progress in Nucleic Acid Research **1**, 135. New York: Acad. Press 1963.
Lucas-Lenard, J., and F. Lipmann: Proc. Nat. Acad. Sci. US **57**, 1050 (1967).

Malkin, L. I., and A. Rich: J. Mol. Biol. **26**, 329 (1967).
Marcker, K. A.: J. Mol. Biol. **14**, 63 (1965).
—, and F. Sanger: J. Mol. Biol. **8**, 835 (1964).
Martin, R. G., D. F. Silbert, D. W. E. Smith, and H. Whitfield: J. Mol. Biol. **21**, 357 (1966).
Matthaei, H., F. Amelunxen, K. Eckert, and G. Heller: Ber. Bunsengesellschaft Phys. Chemie **68**, 735 (1964).
Monro, R. E.: J. Mol. Biol. **26**, 147 (1967).
—, and K. A. Marcker: J. Mol. Biol. **25**, 347 (1967).
—, and D. Vazquez: J. Mol. Biol. **28**, 161 (1967).
Moore, P. B.: J. Mol. Biol. **22**, 145 (1966).
Morgan, A. R., R. D. Wells, and H. G. Khorana: Proc. Nat. Acad. Sci. US **56**, 1899 (1966).

Nakamoto, T., and D. Kolakofsky: Proc. Nat. Acad. Sci. US **55**, 606 (1966).
—, T. W. Conway, J. E. Allende, G. J. Spyrides, and F. Lipmann: Cold Spring Harbor Symposia Quant. Biol. **28**, 227 (1963).
Nathans, D., and F. Lipmann: Proc. Nat. Acad. Sci. US **47**, 497 (1961).
Newton, W. A., J. R. Beckwith, D. Zipser, and S. Brenner: J. Mol. Biol. **14**, 290 (1965).
Nirenberg, M. W., and P. Leder: Science **145**, 1399 (1964).
—, and Matthaei: Proc. Nat. Acad. Sci. US **47**, 1588 (1961).
—, P. Leder, M. Bernfield, R. Brimacombe, J. Trupin, F. Rottman, and C. O'Neal: Proc. Nat. Acad. Sci. US **53**, 1161 (1965).
Nishizuka, Y., and F. Lipmann: (1) Proc. Nat. Acad. Sci. US **55**, 212 (1966).
— — (2) Arch. Biochem. Biophys. **116**, 344 (1966).
Noll, H.: Science **151**, 1241 (1966).
—, T. Staehelin, and F. O. Wettstein: Nature **198**, 632 (1963).
Nomura, M., and C. V. Lowry: Proc. Nat. Acad. Sci. US **58**, 946 (1967).
— —, and C. Guthrie: Proc. Nat. Acad. Sci. US **58**, 1487 (1967).
Notani, G. W., D. L. Engelhardt, W. Konigsberg, and N. Zinder: J. Mol. Biol. **12**, 439 (1965).

Ochoa, S., A. J. Wahba, W. M. Stanley, M. Salas, E. Viñuela, M. A. Smith, M. B. Hille, J. A. Last, and N. A. Elson: Communication at the 7th Intern. Congress of Biochem., Tokyo, Japan. 1967.
Ohta, T., S. Sarkar, and R. E. Thach: Proc. Nat. Acad. Sci. US **58**, 1638 (1967).
Ohtaka, Y., and S. Spiegelman: Science **142**, 493 (1963).

REVEL, M., and F. GROS: Biochem. Biophys. Res. Communs. **27**, 12 (1967).
—, and H. H. HIATT: J. Mol. Biol. **11**, 467 (1965).
RICH, A., J. R. WARNER, and H. M. GOODMAN: Cold Spring Harbor Symposia Quant. Biol. **28**, 269 (1963).
RYCHLÍK, I.: Collection Czechoslov. Chem. Commun. **30**, 2259 (1965).
— (1) Biochim. et Biophys. Acta **114**, 425 (1966).
— (2) Collection Czechoslov. Chem. Commun. **31**, 2583 (1966).
—, and F. ŠORM: Collection Czechoslov. Chem. Commun. **27**, 2433 (1962).
—, S. CHLÁDEK, and J. ZEMLIČKA: Biochim. et Biophys. Acta **138**, 640 (1967).

SALAS, M., M. A. SMITH, W. M. STANLEY, A. J. WAHBA, and S. OCHOA: J. Biol. Chem. **240**, 3988 (1965).
SAMBROOK, J. F., O. P. FAN, and S. BRENNER: Nature **214**, 452 (1967).
SARABHAI, A., and S. BRENNER: J. Mol. Biol. **27**, 145 (1967).
—, A. O. W. STRETTON, S. BRENNER, and A. BOLLE: Nature **201**, 13 (1964).
SCHLESSINGER, D., G. MANGIAROTTI, and D. APIRION: Proc. Nat. Acad. Sci. US **58**, 1782 (1967).
SEEDS, N. W., and T. W. CONWAY: Biochem. Biophys. Res. Communs. **23**, 111 (1966).
SMITH, M. A., M. SALAS, W. M. STANLEY, A. J. WAHBA, and S. OCHOA: Proc. Nat. Acad. Sci. US **55**, 141 (1966).
SPIRIN, A. S.: Ribonucleic acids: Composition, structure and biological role (in Russian), XIX Memorial A.N. Bakh Lecture, April 1963. Izd. Nauka, Moscow. Translation: Macromolecular structure of ribonucleic acids. New York: Reinhold 1964.
— Dokl. Akad. Nauk S.S.S.R. **179**, 1467 (1968); Currents in Modern Biology **2**, 115 (1968).
STRETTON, A. O. W., and S. BRENNER: J. Mol. Biol. **12**, 456 (1965).
SUNDARARAJAN, T., and R. THACH: J. Mol. Biol. **19**, 74 (1966).

TAKANAMI, M.: Biochim. et Biophys. Acta **61**, 432 (1962).
—, and Y. YAN: Proc. Nat. Acad. Sci. US **54**, 1450 (1965).
TERZAGHI, E., Y. OKADA, G. STREISINGER, J. EMRICH, M. INOUYE, and A. TSUGITA: Proc. Nat. Acad. Sci. US **56**, 500 (1966).
— — —, A. TSUGITA, M. INOUYE, and J. EMRICH: Science **150**, 387 (1965).
THACH, R. E., M. A. CECERE, T. A. SUNDARARAJAN, and P. DOTY: Proc. Nat. Acad. Sci. US **54**, 1167 (1965).
—, K. F. DEWEY, and N. MYKOLAJEWYCZ: Proc. Nat. Acad. Sci. US **57**, 1103 (1967).
— —, J. C. BROWN, and P. DOTY: Science **153**, 416 (1966).
TRAUT, R. R., and R. E. MONRO: J. Mol. Biol. **10**, 63 (1964).

WALLER, J. P.: J. Mol. Biol. **7**, 483 (1963).
—, T. ERDÖS, F. LEMOINE, S. GUTTMANN, and E. SANDRIN: Biochim. et Biophys. Acta **119**, 566 (1966).
WARNER, J., M. J. MADDEN, and J. E. DARNELL: Virology **19**, 393 (1963).
—, and A. RICH: Proc. Nat. Acad. Sci. US **51**, 1134 (1964).
WATSON, J. D.: Science **140**, 17 (1963).
— Bull. soc. chim. biol. **46**, 1399 (1964).
— Molecular biology of the gene. New York: Benjamin 1965.
WEBSTER, R. E., D. L. ENGELHARDT, and N. D. ZINDER: Proc. Nat. Acad. Sci. US **55**, 155 (1966).
WEIGERT, M., and A. GAREN: (1) J. Mol. Biol. **12**, 448 (1965).
— — (2) Nature **206**, 992 (1965).
WETTSTEIN, F. O., and H. NOLL: J. Mol. Biol. **11**, 35 (1965).
WILSON, D. A., and I. P. CRAWFORD: J. Biol. Chem. **240**, 4801 (1965).
WILSON, D. B., and D. S. HOGNESS: J. Biol. Chem. **239**, 2469 (1964).

YANOFSKY, C., and J. ITO: J. Mol. Biol. **21**, 313 (1966).

ZABIN, I.: Cold Spring Harbor Symposia Quant. Biol. **28**, 431 (1963).
ZAMIR, A., P. LEDER, and D. ELSON: Proc. Nat. Acad. Sci. US **56**, 1794 (1966).

IV. Appendix: On the Mechanism of the Action of Certain Antibiotics

Allen, D. W., and P. C. Zamecnik: Biochim. et Biophys. Acta **55**, 865 (1962).

Bretscher, M. S., and K. A. Marcker: Nature **211**, 380 (1966).

Cannon, M.: Biochem. J. **104**, 934 (1967).

Connamacher, R. H., and H. G. Mandel: Biochem. Biophys. Res. Communs. **20**, 98 (1965).

Cox, E. C., J. R. White, and J. G. Flaks: Proc. Nat. Acad. Sci. US **51**, 703 (1964).

Das, H., A. Goldstein, and L. Kanner: J. Mol. Pharmacol. **2**, 158 (1966).

Davies, J. E.: Proc. Nat. Acad. Sci. US **51**, 659 (1964).

—, W. Gilbert, and L. Gorini: Proc. Nat. Acad. Sci. US **51**, 883 (1964).

—, L. Gorini, and B. D. Davis: J. Mol. Pharmacol. **1**, 93 (1965).

—, D. S. Jones, and H. G. Khorana: J. Mol. Biol. **18**, 48 (1966).

Erdös, T., and A. Ullmann: Nature **183**, 618 (1959).

— — Nature **185**, 100 (1960).

Flaks, J. G., E. C. Cox, and J. R. White: (1) Biochem. Biophys. Res. Communs. **7**, 385 (1962).

— —, M. L. Witting, and J. R. White: (2) Biochem. Biophys. Res. Communs. **7**, 390 (1962).

Franklin, T. J.: Biochem. J. **87**, 449 (1963).

Gilbert, W.: J. Mol. Biol. **6**, 389 (1963).

Hancock, R.: Biochem. J. **78**, 7 P. (1961).

Hierowski, M.: Proc. Nat. Acad. Sci. US **53**, 594 (1965).

—, and Z. Kurylo-Borowska: Biochim. et Biophys. Acta **95**, 578 (1965).

Julian, G. R.: J. Mol. Biol. **12**, 9 (1965).

Kaji, H.: Biochim. et Biophys. Acta **134**, 134 (1967).

—, and A. Kaji: Proc. Nat. Acad. Sci. US **54**, 213 (1965).

—, I. Suzuka, and A. Kaji: J. Biol. Chem. **241**, 1251 (1966).

Kućan, Z., and F. Lipmann: J. Biol. Chem. **239**, 516 (1964).

Kurylo-Borowska, Z., and M. Hierowski: Biochim. et Biophys. Acta **95**, 590 (1965).

Mager, I., M. Benedict, and M. Artman: Biochim. et Biophys. Acta **62**, 202 (1962).

Monro, R. E.: J. Mol. Biol. **26**, 147 (1967).

—, and K. A. Marcker: J. Mol. Biol. **25**, 347 (1967).

—, and D. Vazquez: J. Mol. Biol. **28**, 161 (1967).

Morris, A., and R. Schweet: Biochim. et Biophys. Acta **47**, 415 (1961).

Nakamoto, T., T. W. Conway, J. E. Allende, G. J. Spyrides, and F. Lipmann: Cold Spring Harbor Symposia Quant. Biol. **28**, 227 (1963).

Nathans, D.: Proc. Nat. Acad. Sci. US **51**, 585 (1964).

—, and F. Lipmann: Proc. Nat. Acad. Sci. US **47**, 497 (1961).

—, G. von Ehrenstein, R. Monro, and F. Lipmann: Fed. Proc. **21**, 127 (1962).

Pestka, S.: J. Biol. Chem. **241**, 367 (1966).

—, and M. W. Nirenberg: J. Mol. Biol. **21**, 145 (1966).

—, R. Marshall, and M. Nirenberg: Proc. Nat. Acad. Sci. US **53**, 639 (1965).

Rendi, R., and S. Ochoa: J. Biol. Chem. **237**, 3711 (1962).

Rychlík, I.: (1) Biochim. et Biophys. Acta **114**, 425 (1966).

— (2) Collection Czechoslov. Chem. Commun. **31**, 2583 (1966).

Smith, J. D., R. R. Traut, G. M. Blackburn, and R. E. Monro: J. Mol. Biol. **13**, 617 (1965).

Speyer, J. F., P. Lengyel, and C. Basilio: Proc. Nat. Acad. Sci. US **48**, 684 (1962).

— — —, A. J. Wahba, R. S. Gardner, and S. Ochoa: Cold Spring Harbor Symposia Quant. Biol. **28**, 559 (1963).

Spotts, C. R., and R. Y. Stanier: Nature **192**, 633 (1961).

Suarez, G., and D. Nathans: Biochem. Biophys. Res. Communs. **18**, 743 (1965).

SUZUKA, I., H. KAJI, and A. KAJI: Proc. Nat. Acad. Sci. US **55**, 1483 (1966).

TANAKO, K., and H. TERAOKA: Biochim. et Biophys. Acta **113**, 204 (1966).
TAUBMAN, S., N. JONES, F. YOUNG, and J. CORCORAN: Biochim. et Biophys. Acta **123**, 438 (1966).
—, F. E. YOUNG, and J. W. CORCORAN: Proc. Nat. Acad. Sci. US **50**, 955 (1963).
TRAUT, R. R., and R. E. MONRO: J. Mol. Biol. **10**, 63 (1964).

VAZQUEZ, D.: Biochem. Biophys. Res. Communs. **12**, 409 (1963).
— Biochem. Biophys. Res. Communs. **15**, 464 (1964).
— (1) Biochim. et Biophys. Acta **114**, 277 (1966).
— (2) Biochim. et Biophys. Acta **114**, 289 (1966).
—, and R. E. MONRO: Biochim. et Biophys. Acta **142**, 155 (1967).

WEBER, M., and J. DEMOSS: Proc. Nat. Acad. Sci. US **55**, 1224 (1966).
WOLFE, A. D., and F. E. HAHN: Science **143**, 1445 (1964).
— — Biochim. et Biophys. Acta **95**, 146 (1965).

YARMOLINSKY, M. B., and G. L. HABA: Proc. Nat. Acad. Sci. US **45**, 1721 (1959).

ZAMIR, A., P. LEDER, and D. ELSON: Proc. Nat. Acad. Sci. US **56**, 1794 (1966).

Subject Index

Typesetting and printing: Carl Ritter & Co., Wiesbaden

Molecular Biology, Biochemistry and Biophysics
Molekularbiologie, Biochemie und Biophysik

Volumes published:

Vol. 1 J. H. van't Hoff: Imagination in Science. Translated into English, with Notes and a General Introduction by G. F. Springer. With 1 portrait. VI, 18 pages. 1967. Soft cover DM 6,60 / US $1.65

Vol. 2 K. Freudenberg and A. C. Neish: Constitution and Biosynthesis of Lignin. With 10 figures. IX, 129 pages. 1968. Bound DM 28,— / US $ 7.—

Vol. 3 T. Robinson: The Biochemistry of Alkaloids. With 37 figures. X, 149 pages. 1968. Bound DM 39,—/US $ 9.75

Vol. 4 A. S. Spirin and L. P. Gavrilova: The Ribosome. With 26 figures. X, 161 pages. 1969. Bound DM 54,—/US $ 13.50

Vol. 5 B. Jirgensons: Optical Rotatory Dispersion of Proteins and Other Macromolecules. With 65 figures. XI, 166 pages. 1969. Bound DM 46,—/US $ 11.50

Vol. 6 F. Egami and K. Nakamura: Microbial Ribonucleases. With 5 figures. IX, 90 pages. 1969. Bound DM 28,—/US $ 7.—

Forthcoming volume:

Vol. 7 F. Hawkes: Nucleic Acids and Cytology. With approx. 35 figures. Approx. 288 pages. Due November 1969. Bound DM 58,— / US $ 14.50

Volumes being planned:

Abeles, R. H., and W. S. Beck: The Mechanism of Action of (Cobamide) Vitamin B_{12} Coenzymes

Aida, Kô: Amino Acid Fermentation

Ando, T.: Protamines

Anker, H. S.: Studies on the Mechanism and Control of Fatty Acid Biosynthesis

Antonini, E.: Kinetic Structures and Mechanism of Hemoproteins

Beermann, W.: Puffs and their Function in Developmental Biology

Breuer, H., and K. Dahm: Mode of Action of Estrogens

Brill, A. S.: The Role of Heavy Metals in Biological Oxidation Processes

Burns, R. C., and R. W. F. Hardy: Nitrogen Fixation in Bacteria and Higher Plants

Callan, H. G.: Lampbrush Chromosomes

Chapman, D.: Molecular Biological Aspects of Lipids in Membrane

Chichester, C. O., H. Yokoyama and T. O. M. Nakayama: Biosynthesis of Carotenoids

Dumonde, D. C.: Cell and Tissue Antigens

Dus, K. M.: Analysis of Proteins

Eley, D. D.: Biological Materials as Semi-Conductors

Fender, D. H.: Problems of Visual Perceptions

Gibson, I.: Paramecium Aurelia and Cytoplasmic Factors

Volumes being planned:

GIBSON, Q. H.: Rapid Reactions in Biochemistry

GRECZ, N.: Molecular Radiation Biophysics

GRUBB, R.: Genetical and Medical Aspects of Immunoglobulin Differentiation

GUILLEMIN, R.: Hypothalamic Releasing Factors

HAUGAARD, N., and E. HAUGAARD: Mechanism of Action of Insulin

HEINZ, E., and W. WILBRANDT: Biological Transports

HOFFMANN-OSTENHOF, O.: Biochemistry of the Inositols

HOFMANN, TH.: Chymotrypsin and other Serine Proteinases

JARDETZKY, O.: Nuclear Magnetic Resonance in Molecular Biology

JOLLÈS, P., and A. PARAF: Biological and Chemical Basis of Adjuvants

KLEINSCHMIDT, A. K., and D. LANG: A Manual for Electron Microscopy of Nucleic Acid Molecules

KLENK, E.: Cerebrosides, Gangliosides and other Glycolipids

LARNER, J.: Chemistry of Hereditary Disease

LUDWICK, T. M., J. E. GANDER and R. JENNESS: Biosynthesis of Milk

LUKENS, R. J.: Chemistry of Fungicides

MATHEWS, M. B.: Molecular Evolution of Connective Tissue

MATTHAEI, J. H.: Protein Biosynthesis

MAURER, P. H.: Chemical and Structural Basis of Antigenicity

McMANUS, T. J.: Red Cell Ion Transport and Related Metabolic Processes

MEIENHOFER, J.: Synthesis of Biologically Active Peptides

MIDDLETON, J. T., and J. B. MUDD: Air Pollution

NELSON, R. A.: The Complement System. Physicochemical Characterization and Biologic Reactivities

OKUNUKI, K.: Cytochromes

OSBORN, M. J., and L. ROTHFIELD: Bacterial Cell Wall Synthesis and Function

PAUL, K. G.: Peroxidases and Catalases

PORATH, J.: Modern Methods of Separation of Proteins and other Macromolecules

QUASTEL, J. H.: Chemical Control of Cell Metabolism and Function

REEVES, P.: Bacteriocins

RODWELL, V. W., and R. SOMERVILLE: Metabolism of Amino Acids

RUNNSTRÖM, J.: Molecular Biology of the Fertilization Process

SCHANNE, O., and ELENA RUIZ DE CERETTI: Impedance Measurements in Biological Cells

SELIGER, H. H.: Bioluminescence

SHULMAN, S., and F. MILGROM: Organ Specificity

STADTMAN, E. R.: Enzymatic Regulation of Metabolism

Ts'o, O. P., and P. A. STRAAT: Nucleic Acids Polymerases

VINNIKOV, J. A.: Principles of Structural, Chemical and Functional Organization of Sensory Receptors

WERLE, E.: Plasma Kinines

YAGI, K.: Biochemistry of Flavins

ZWEIFACH, B. W.: Shock